Electromagnetic Field in Advancing Science and Technology

Edited by Hai-Zhi Song, Kim Ho Yeap and Magdalene Wan Ching Goh

Published in London, United Kingdom

Electromagnetic Field in Advancing Science and Technology
http://dx.doi.org/10.5772/intechopen.100711
Edited by Hai-Zhi Song, Kim Ho Yeap and Magdalene Wan Ching Goh

Contributors
Ramesh Mohan Thamankar, Vithaldas Vallabhdas Raja, Jun-ichi Sugiyama, Hayato Sugiyama, Chika Sato, Maki Morizumi, Kim Ho Yeap, Len Luet Ng, Magdalene Wan Ching Goh, Veerendra Dakulagi, Salama AbdelHady, Cancio Monteiro, Rafael Zamorano Ulloa, Kostadin Trenčevski

First published in London, United Kingdom, 2023 by IntechOpen
IntechOpen is the global imprint of INTECHOPEN LIMITED, registered in England and Wales, registration number: 11086078, 5 Princes Gate Court, London, SW7 2QJ, United Kingdom

British Library Cataloguing-in-Publication Data
A catalogue record for this book is available from the British Library

Additional hard and PDF copies can be obtained from orders@intechopen.com

Electromagnetic Field in Advancing Science and Technology
Edited by Hai-Zhi Song, Kim Ho Yeap and Magdalene Wan Ching Goh
p. cm.
Print ISBN 978-1-80356-677-1
Online ISBN 978-1-80356-678-8
eBook (PDF) ISBN 978-1-80356-679-5

Meet the editors

Hai-Zhi Song is a professor at the Southwest Institute of Technical Physics and the University of Electronic Science and Technology, China. He obtained his Ph.D. degree from Peking University in 1995 and has worked at Nanjing University, Catholic University Leuven, Tsukuba University, Fujitsu Lab. Ltd., and Tokyo University. His research interests are optoelectronics, nano-materials, and quantum information processing. His achievements include: preparing materials and devices for the silicon-based light emitter, quantum dot/well laser, and single-photon source/detector; inventing techniques to control the size and site of quantum dots; developing focal-plane single-photon detectors; designing a series of 2D metal-halides as potential optoelectronic and catalysis materials; and realizing quantum devices such as fiber-integrated photon-pair-entangler, heralded single photon emitter and fiber quantum memories.

Kim Ho Yeap is an associate professor at Universiti Tunku Abdul Rahman, Malaysia. He received his BEng (Hons) in electrical and electronics engineering from Universiti Teknologi Petronas in 2004, MSc in microelectronics from Universiti Kebangsaan Malaysia in 2005, and Ph.D. from Universiti Tunku Abdul Rahman in 2011. He conducted research at the University of Oxford, UK in 2008 and Nippon Institute of Technology, Japan in 2015. He is a senior member of the IEEE, a Professional Engineer registered with the Board of Engineers, Malaysia, a Chartered Engineer registered with the UK Engineering Council, and an ASEAN Chartered Professional Engineer. He is an external examiner and external course assessor for Wawasan Open University. He has received 23 research grants and an award for university teaching excellence. He has published more than 100 research articles, including patents, refereed journal papers, conference proceedings, books, and book chapters. From 2017 to 2022, he was the Editor in Chief of the *i-manager's Journal on Digital Signal Processing*. He has also been a guest editor for the *Journal of Applied Environmental and Biological Sciences* and the *Journal of Fundamental and Applied Sciences*.

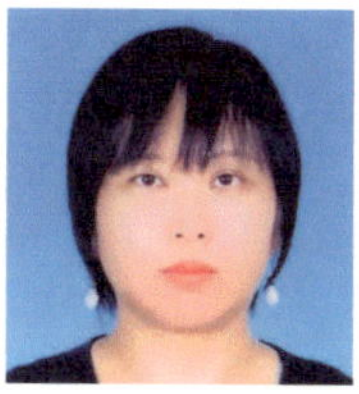

Dr. Magdalene Goh obtained her diploma in electrical and electronics engineering from Inti College before leaving for the UK to pursue her BEng in electrical engineering and electronics and, later on, her Ph.D. She had previously worked in industry in the areas of semiconductor process technology, silicon wafer characterizations, mask layout design, analog circuits design and design for testability (DFT). Since 2014 she has been actively involved with the radio telescope team at Penang Science Cluster, where she works with a team of volunteers from both academia and industry to create radio astronomy curricula for school pupils and college students. Dr. Goh's research interests are in semiconductor physics and electromagnetics.

Contents

Preface

The discovery of the electromagnetic field is one of the greatest scientific achievements of the 19th century. As one of the most important physical phenomena, the electromagnetic field extends into most human activities. It is also a fundamental concept in modern science and technology, with impacts on manufacturing, medicine, astronomy, meteorology, and many other fields.

In recent decades, there has been rapid development of many new areas of science and technology including information and communications technology, microelectronics, optics, and molecular biology. Electromagnetic field theory has been applied in such areas as microwave communication, remote sensing, navigation, photon detection and imaging, biomedicine, aerospace technology, weather forecasting, and so on. Civilian applications are mainly concentrated in such areas as mobile terminals, wireless communications, and radio frequency identification technology, while military applications involve national security systems and defense equipment, such as radar, navigation, satellites, etc. Wherever communication is required (on land, under water, in the sky or in space), advanced electromagnetic theory and technology are needed. Electromagnetic theory and engineering face increasing challenges with the development of many new fields such as millimeter wave and sub-millimeter wave technologies, microwave monolithic integrated circuits, smart antennas, portable devices, new artificial electromagnetic materials, computational electromagnetics, electromagnetic compatibility, terahertz technology and bio-electromagnetics.

This book reviews recent achievements in the theory and application of electromagnetism in highly developed sciences and technologies. Three broad aspects are considered: new insights into the electromagnetic field; its application in scientific research; and its application in technology. Chapter 1 offers new ideas about the electromagnetic field in 3-dimensional space. In electromagnetic interactions, the magnetic field of the spin and the electrostatic field of charges are studied separately. Gravitation is also considered to emphasize the similarity between it and electromagnetism. Chapter 2 explores the impacts of the new definitions of electric current and magnetic flux on the explanation of electromagnetic and photoelectric phenomena; also on the proper understanding of the duality confusion and wireless power transfer history. As the electric charge and magnetic flux are recognized as forms of energy, the ampere can be removed from the SI system. Chapter 3 considers electric fields within lightning clouds in atmospheres of different properties. A new concept, electric field at the verge of discharge, is introduced, which might be the result of two specific charge configurations. By focusing on observation points close to lightning clouds, valuable information can be extracted from the structures that produce pronounced sinks/sources with large divergences of the electric field. Chapter 4 discusses the dielectric constant of materials in microwave electromagnetic fields, extending Debye relaxation to explain material behaviors. The chapter also describes which form should be taken if there is a non-thermal effect upon chemical reactions under microwave radiation. Chapter 5 explains two types of electromagnetic power consumption found in a complementary metal oxide semiconductor (CMOS) circuit. Since the dynamic power consumption is usually significantly higher than its static counterpart, some power-saving techniques are suggested, such as reducing the supply voltage, clock frequency, clock power, and dynamic effective capacitance. Chapter 6 presents a low-power adiabatic CMOS/FinFETs circuit for low-power secure logic application. The performance of the proposed circuits is compared with that of other circuit topologies available in the literature in terms of speed, power consumption and other metrics. The proposed method is proved to be effective in optimizing CMOS/FinFETs. Chapter 7 reviews the effects of various external stimuli, including electromagnetic fields, on the characteristics of resistive switching devices. Different material systems used to manufacture memristors are discussed in detail. 2D materials have unique electrical, optical and mechanical properties over a wide electromagnetic spectral range, opening up opportunities for the development of new and efficient devices working with electromagnetic fields.

We hope that the publication of this book heralds a bright future for electromagnetic science and technology. We sincerely thank all the authors for their valuable contributions, and for their commitment to the timely completion and updating of the chapters. We would particularly like to thank Sara Debeuc, Author Service Manager at IntechOpen, for her excellent organization and her coordination work between the authors, without which publication of this book could not have been achieved.

With the publishing of this book, we wish for a bright future for electromagnetic science and technology.

Hai-Zhi Song
Professor, Principal Investigator,
Southwest Institute of Technical Physics,
Chengdu, China

Professor,
University of Electronic Science and Technology of China,
Chengdu, China

Kim Ho Yeap
Universiti Tunku Abdul Rahman,
Kampar, Perak, Malaysia

Magdalene Wan Ching Goh
Wawasan Open University,
Penang, Malaysia

Chapter 1

Geometrical Aspects of the Electromagnetic Field

Kostadin Trenčevski

Abstract

The space–time is based on space as three-dimensional sphere, space rotations also as three-dimensional sphere and time which is homeomorphic to the Euclidean three-dimensional space. There are four basic exchanges among them and four induced exchanges, which lead to the basic interactions in the nature. This geometrical approach enables to obtain new viewpoint especially on the basic interactions and their geometrical interpretations. More attention is devoted to the electromagnetic interactions, where the magnetic field of the spinning bodies is studied separately from the electromagnetic field of the charged bodies. It is also considered the gravitational interaction in order to emphasize the similarities between them and the properties which separate them.

Keywords: Lie groups, electric field, magnetic field, rotation, time

1. Introduction

The space and time were subject of interest from the old civilizations up to the present time. They were separated many centuries ago. Even in the Newton theory they are still separated and the time flow was considered as uniform phenomena in the universe. Remarkable approach in understanding the space and time was done by the well known scientist and philosopher Roger Boscovich (1711–1787), who was not well understood at that time. He made distinction between the real space–time and the space–time according to our observations (Ref. [1]). More than one century before the Special Relativity, he wrote that there does not exist an absolute space in rest, i.e. about the relativity among the moving systems.

Basic argument in the Spacial Relativity is the assumption that there does not exist strong distinction between the space and time, which is obvious from the Lorentz transformations. In the recent references this idea was generalized between the space (S), space rotations (SR) and the time (T) [2–6]. For each small body besides its three spatial coordinates, can be jointed also three degrees of freedom about its rotation in the space and also three degrees of freedom for the velocity of the considered body. The basic assumption is that these six degrees of freedom are of the same importance as the basic three spatial coordinates and so they can be considered as dimensions. The spaces S and SR are both homeomorphic to S^3 and can be considered as the group of quaternions with module 1, which is locally isomorphic to $SO(3, \mathbb{R})$. Note that if the space SR is not admitted, we would not be able to turn. The time T is homeomorphic to $\mathbb{R}^3$, and if it is not admitted, everything would be static. Each two of these three sets may interfere analogously to the space and time in the Special Relativity.

It is deduced in Refs. [2, 3] that the general form of three-dimensional time of one point is given by

$$\mathbf{n}t + \frac{\mathbf{v} \times \mathbf{r}}{c^2}, \tag{1}$$

where t is the 1-parameter time, $\mathbf{n}$ is the unit vector of its velocity, $\mathbf{v}$ is its velocity and $\mathbf{r}$ is the radius vector of the considered point. The last term gives possibility for a new view of the quantum mechanics, where the quantization of the angular momentum is indeed quantization of the three-dimensional time.

The group of Lorentz transformations $O_{+}^{\uparrow}(1, 3)$ is known to be isomorphic to the complex Lie group $SO(3, \mathbb{C})$. If we consider this complex group as a Lie group of real 6×6 matrices, then that group is the required Lie group which connects the spaces S and T. This Lie group of 6×6 transformations has Lie algebra consisting of matrices of type

$$\begin{bmatrix} X & Y \\ -Y & X \end{bmatrix}, \tag{2}$$

where X and Y are antisymmetric 3×3 matrices. This temporal Lie group is denoted by G_t. Moreover, in Ref. [4] the Lorentz transformations are converted as transformations $S \times T \to S \times T$ via the Lie group G_t.

The Lie group which connects the spaces S and SR has Lie algebra which consists of matrices of type

$$\begin{bmatrix} X & Y \\ Y & X \end{bmatrix}, \tag{3}$$

where X and Y are antisymmetric 3×3 matrices. This Lie group is generated by the following six matrices of type

$$\begin{bmatrix} 1 & 0 & 0 & 0 & 0 & 0 \\ 0 & \cos\psi & 0 & 0 & 0 & \sin\psi \\ 0 & 0 & \cos\psi & 0 & -\sin\psi & 0 \\ 0 & 0 & 0 & 1 & 0 & 0 \\ 0 & 0 & \sin\psi & 0 & \cos\psi & 0 \\ 0 & -\sin\psi & 0 & 0 & 0 & \cos\psi \end{bmatrix},$$

(translation along the x-axis)

$$\begin{bmatrix} \cos\psi & 0 & 0 & 0 & 0 & -\sin\psi \\ 0 & 1 & 0 & 0 & 0 & 0 \\ 0 & 0 & \cos\psi & \sin\psi & 0 & 0 \\ 0 & 0 & -\sin\psi & \cos\psi & 0 & 0 \\ 0 & 0 & 0 & 0 & 1 & 0 \\ \sin\psi & 0 & 0 & 0 & 0 & \cos\psi \end{bmatrix},$$

(translation along the y-axis)

$$\begin{bmatrix} \cos\psi & 0 & 0 & 0 & \sin\psi & 0 \\ 0 & \cos\psi & 0 & -\sin\psi & 0 & 0 \\ 0 & 0 & 1 & 0 & 0 & 0 \\ 0 & \sin\psi & 0 & \cos\psi & 0 & 0 \\ -\sin\psi & 0 & 0 & 0 & \cos\psi & 0 \\ 0 & 0 & 0 & 0 & 0 & 1 \end{bmatrix}, \tag{4}$$

(translation along the z-axis)

$$\begin{bmatrix} 1 & 0 & 0 & 0 & 0 & 0 \\ 0 & \cos\psi & \sin\psi & 0 & 0 & 0 \\ 0 & -\sin\psi & \cos\psi & 0 & 0 & 0 \\ 0 & 0 & 0 & 1 & 0 & 0 \\ 0 & 0 & 0 & 0 & \cos\psi & \sin\psi \\ 0 & 0 & 0 & 0 & -\sin\psi & \cos\psi \end{bmatrix},$$

(rotation around the x-axis)

$$\begin{bmatrix} \cos\psi & 0 & -\sin\psi & 0 & 0 & 0 \\ 0 & 1 & 0 & 0 & 0 & 0 \\ \sin\psi & 0 & \cos\psi & 0 & 0 & 0 \\ 0 & 0 & 0 & \cos\psi & 0 & -\sin\psi \\ 0 & 0 & 0 & 0 & 1 & 0 \\ 0 & 0 & 0 & \sin\psi & 0 & \cos\psi \end{bmatrix},$$

(rotation around the y-axis)

$$\begin{bmatrix} \cos\psi & \sin\psi & 0 & 0 & 0 & 0 \\ -\sin\psi & \cos\psi & 0 & 0 & 0 & 0 \\ 0 & 0 & 1 & 0 & 0 & 0 \\ 0 & 0 & 0 & \cos\psi & \sin\psi & 0 \\ 0 & 0 & 0 & -\sin\psi & \cos\psi & 0 \\ 0 & 0 & 0 & 0 & 0 & 1 \end{bmatrix}, \tag{5}$$

(rotation around the z-axis),

where ψ is an arbitrary parameter. This is space group which connects the spaces S and SR and is denoted by G_S. The group is isomorphic to the spin group Spin(4) [7].

Analogously as the speed of light is a coefficient of proportionality between the length and the time, in case of the spaces S and SR we have the following situation. The elements of the space S are measured in length while elements of spatial rotations SR are measured in angles. Consequently there exists a local constant as a coefficient of proportionality between S and SR. This local constant is called *radius of range* R. It depends mainly of the mass of the particles, galaxies and the universe. Now the connection between SR and T is determined, and the group which connects the spaces

SR and T is the temporal group G_t. The radius of range is of the same importance as the velocity of light, which is a coefficient of proportionality between the spaces S and T. While the Lorentz transformations tend to Galilean transformations for small velocities, if the radius of range is close to infinity, then the group G_s tends to the affine group of translations and rotations in the Euclidean space.

The three-dimensional time was investigated also by another authors, for example in Refs. [8–15].

2. Basic four exchanges among the spaces *S*, *SR* and *T*

The spaces S, SR and T are not separated. Analogously as there does not exist a strong distinction between the space and time in the Special Relativity, there exist similar exchanges in this case. In [16] are given four basic exchanges and four generalizations of them. Let the elements of the spaces S, SR and T are denoted respectively by s, r and t. The basic four cases are presented on **Figure 1** and they are

$$1.\ r \to s, \quad 2.\ s \to r, \quad 3.\ r \to t, \quad 4.\ s \to t. \tag{6}$$

Here the notation $x \to y$ denotes that if there is a constraint of $x \in X$, then it converts into $y \in Y$. The composite exchanges $s \to r \to t$ and $r \to s \to t$ are also admitted, while the exchange of type $x \to y \to x$ is not admitted because we may not return back to x, if x was constrained previously. The exchanges $t \to s$ and $t \to r$ are not admitted, because it is impossible to constraint the time. For example, we can not conduct the speed of time without constraints of S and SR. Since the exchange $s \to r \to s$ is not possible, the case 3. must be of type $s \to r \to t$, and since the exchange $r \to s \to r$ is not possible, the case 4. must be of type $r \to s \to t$.

The second case ($s \to r$) means that if the space displacement is constrained (completely or partially), then it induces a spatial rotation. It can be interpreted if we consider a rigid body, which moves with a velocity $\mathbf{v}$. Assume that one point O of the rigid body is constrained to move. Then the body will start to rotate around the point O. Each point P of the body intends to rotate with angular velocity $\mathbf{w}_P$. This means that the case $s \to r$ occurs. However, the rigid body will rotate globally with an averaged angular velocity $<\mathbf{w}>$, and it depends on the distribution of the mass. So at the point P the angular velocity $\mathbf{w}_P - <\mathbf{w}>$ is constrained. It will cause a change of the speed of time if $\mathbf{w}_P - <\mathbf{w}> \neq 0$, and this is the third case $r \to t$.

The first case $r \to s$ means that if the rotation is not permitted, then the particles will be displaced, which is the induced spin motion. This induced spin velocity, or shortly spin velocity, was considered in some previous papers, for example in [17]. Let a solid body moves including some rotations, then each particle intends to rotate according to its spatial trajectory. But that rotation is often constrained (completely or

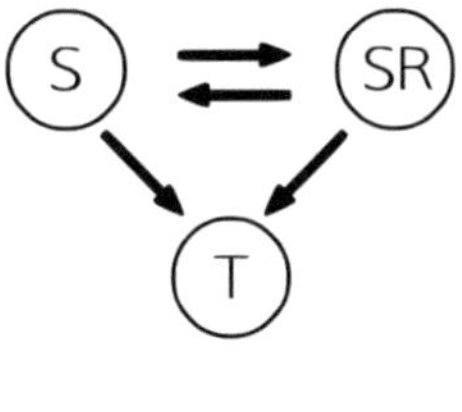

Figure 1.
Spaces S, SR and T and the possible exchanges between them.

partially), because we consider a solid body, and the particles of the body do not have freedom to rotate as they intend to rotate. So, the constrained space rotation induces the spin velocity. This is the reason for circular motion of the spinning bodies when they rotate on a horizontal plane or in free fall motion. Moreover, the spin velocity of the Earth is considered in Ref. [17], and as a consequence the change of the Earth's angular velocity with period of 6 months was deduced with accuracy 5%, compared with the measurements. In the recent Ref. [6] the spin velocity is determined via the curvature and torsion of the trajectory of the considered particle.

The spin velocity is non-inertial and it will be denoted by capital letter **V** in order to distinguish from the classical inertial velocity **v**. The main property of the inertial velocities is that temporal axes are inclined, i.e. no-parallel, analogously as the temporal axes are mutually non-parallel in case of two mutually moving inertial coordinate systems. It is also important that if a coil moves with spin velocity in magnetic field, there will not appear an electromotive force, because for electromotive force it is necessary to have inertial velocity between the magnetic field and the coil. Indeed, the spin velocity means simply displacement in the space and there does not appear change of speed of time by the coefficient $\sqrt{1-V^2/c^2}$. But, if the spin velocity is constrained completely or partially, then the constrained part converts into inertial velocity $\mathbf{v} = -\mathbf{V}/\sqrt{1+V^2/c^2}$, which is indeed the fourth case $s \to t$. This velocity **v** can be non-observable, but it is important that the speed of time is changed at that position. Then the coil in magnetic field will produce electromotive force via the inertial velocity.

We give now another important examples of the case 4.

Let a non-rotating small body initially rests with respect to the Earth on infinite distance. When the body comes at the surface of the Earth, it is not permitted to go further. This constraint causes time displacement, such that the time will be slower. More precisely, if the velocity at the surface is equal to v, then the constraint for the space displacement will induce slower time for coefficient $\lambda = \sqrt{1-\frac{v^2}{c^2}}$ and this is the case 4 ($s \to t$). Using that $v \approx \sqrt{2GM/R}$, the time on the surface of the Earth is slower for coefficient $\lambda \approx \sqrt{1-\frac{2GM}{Rc^2}} \approx 1-\frac{GM}{Rc^2}$, which is also well known from the General Relativity up to approximation of c^{-2}.

In this example it was not important that the particle moved under gravitation. So let us consider now two charged bodies with charges e_1 and e_2, masses m_1 and m_2, and $m_2 << m_1$. Assume that the second body is not rotating and emits waves with frequency ν_0 on infinity distance (**Figure 2**). If the second body is placed close to the first body on distance r between their centers and assume that the distance r remains a constant. In case of free fall motion (if $e_1e_2 < 0$) the velocity of the second body on distance r would be $\sqrt{\frac{-e_1e_2}{2\pi\varepsilon_0 r m_2}}$. Then according to the case 4, analogously to the gravitational case now we have slower speed of time and the new frequency is given by

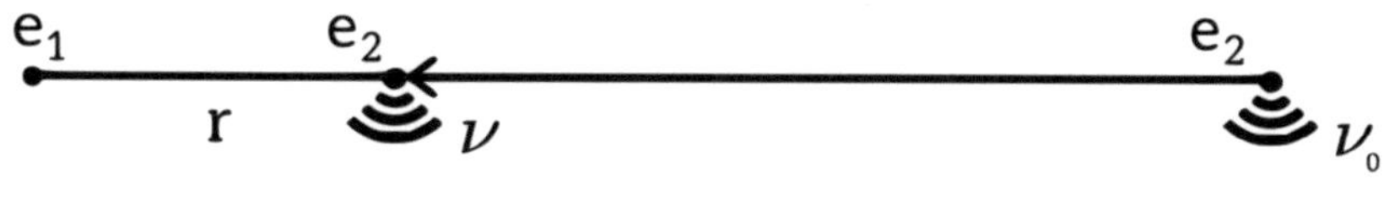

Figure 2.
The frequency ν changes when the distance r changes.

$$\nu = \nu_0 \left(1 + \frac{e_1 e_2}{4\pi\varepsilon_0 r m_2 c^2}\right). \tag{7}$$

This formula is true also in case $e_1 e_2 > 0$.

3. The induced (extended) four exchanges among the spaces *S*, *SR* and *T*

The basic cases 1-4 describe the changes of the positions, rotations and temporal coordinates of a chosen particle in case of some constraint. The induced four exchanges, will be denoted by 1*, 2*, 3*, and 4* and they are analogous to the cases 1-4 respectively. In cases 1*-4* we need to describe the fields which appear in case of constraints and to describe the interactions between the particles. Now the basic role play the group structures, especially their non-commutativity. As a consequence some interactions appear in these cases. The non-commutativity leads to some angles. They are functions of the spatial coordinates, and so instead of differentiation by t we should apply the operator rot. Half of this quantity is admitted, and half is not admitted. That half which is not admitted will be converted into acceleration by multiplication by a constant K. This principle of admitted and unadmitted angular velocities is general and further in each section it will be used. It is shown and supported by the following example.

Example 1. Let us consider the vector field $\mathbf{v}(x, y, z) = (-yw, xw, 0)$. Since $\text{rot}\mathbf{v} = (0,0,2w)$, obviously half of it is admitted as angular velocity $\mathbf{w} = (0, 0, w)$, and the other half which in not admitted converts into space displacement, which is observed as centripetal acceleration $(a_x, a_y, a_z) = \mathbf{v} \times (-\mathbf{w})$. These two effects do not occur simultaneously, but they change permanently in a very short interval Δt as a process of discretization, because these two effects are mutually dependent: in the first time interval Δt the body rotates around its axis with angular velocity $2w$, in the next interval Δt it falls toward the center with centripetal acceleration $\mathbf{v} \times (-2\mathbf{w})$, and it periodically repeats. So, we observe that the average angular velocity is $(0, 0, w)$, and the average centripetal acceleration is $\mathbf{v} \times (-\mathbf{w})$.

In the cases 1* and 2* we should not use the Lorentz transformation and should not use the operator $\partial/\partial t$. The time exists in these cases, but caused by another external reasons and it should not be actively used. The motions in these two case are analogous to the induced spin motions.

The cases 3* and 4* lead to the electromagnetic and gravitational interactions. They are more sophisticated interactions and will be considered in more details in the next sections. It is convenient to consider the gravitation together with the electromagnetic

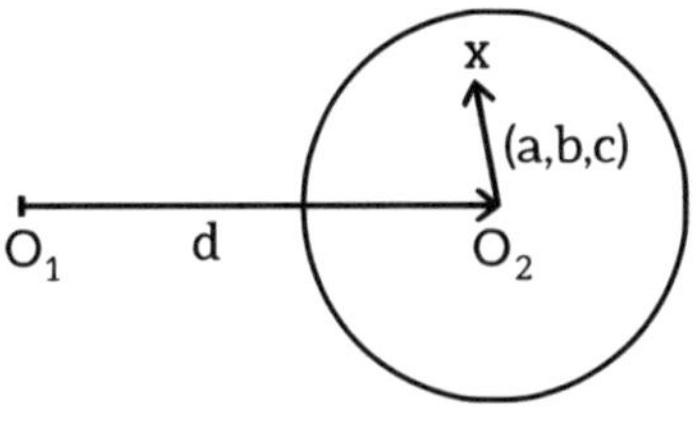

Figure 3.
The strong interaction is a consequence of non-commutativity of translations for the vectors $\mathbf{d}$ *and* (a, b, c) *in* $S \times SR$.

interaction because we need to emphasize the similarity and also where these two interactions differ. To the end of this section we give a brief view of the case 1*.

Let us consider two nucleons with masses m_1 and m_2, momentums of inertia I_1 and I_2, radiuses of range R_1 and R_2, and centers at O_1 and O_2 (**Figure 3**). If X is arbitrary point of the second nucleon, then the composition of the translations for vectors $\mathbf{d} = \overrightarrow{O_1O_2}$ and $\overrightarrow{O_2X} = (a, b, c)$ is non-commutative in the Lie group G_s and contains also a rotation for an angle. The operator $\frac{1}{2}$rot should be applied and yields [16]

$$\frac{1}{2}\mathrm{rot}\varphi = -\frac{1}{2}\sin\frac{d}{R_1}\cdot\sin\frac{d}{R_2}\cdot\frac{R_1R_2}{d^2}\frac{\mathbf{d}}{d^2}. \tag{8}$$

Assume that the two nucleons have opposite spins, and assume that the rotation is constrained (or not admitted). Then the space displacement appears as a force, if we multiply the previous value by the constant μv_0^2, where $\mu = m_1m_2/(m_1+m_2)$ is the reduced mass, which appears in binary systems. The velocity v_0 is a local constant, analogously as the radius of range is a local constant. In case of the nucleons, this constant is close to c, but at this moment we are unable to prove it. In the next sections about the cases 3* and 4*, where the time has role, the corresponding constant takes value c. The force of the second nucleon toward the first nucleon is equal to

$$\mathbf{f}_{2,1} = -\frac{m_1m_2v_0^2}{2(m_1+m_2)}\sin\frac{d}{R_1}\cdot\sin\frac{d}{R_2}\cdot\frac{R_1R_2}{d^3}\frac{\mathbf{d}}{d}. \tag{9}$$

Assume that the space displacement is constrained (or not admitted). Then it appears rotation of the two nucleons. The admitted value (8) induces that the two nucleons also have opposite angular momentums and we determine now the magnitude of this angular momentum. This can be obtained by multiplication with $I_1I_2v_0/(I_1+I_2)$ and hence the required angular momentum is equal to

$$\mathcal{L} = \pm\frac{I_1I_2v_0}{2(I_1+I_2)}\sin\frac{d}{R_1}\cdot\sin\frac{d}{R_2}\cdot\frac{R_1R_2}{d^3}\frac{\mathbf{d}}{d}. \tag{10}$$

The force (9) can be applied for two galaxies and then it avoids collisions between them, because the forces appear to be also repulsive on some distances.

The following special case, where $m_2/m_1 \approx 0$, $I_2/I_1 \approx 0$ and $R_2 \approx \infty$, is of special interest. It means that the second body is much smaller and can be considered as a test particle. The first body can be arbitrary, for example it can be the massive body at the center of a galaxy, while the test body can be arbitrary star. If the angular velocity is constrained, the required space displacement is observed by the acceleration

$$\mathbf{a} = -\sin\frac{d}{R}\cdot\frac{v_0^2R}{2d^2}\frac{\mathbf{d}}{d}, \tag{11}$$

where the local constant v_0 is much smaller compared with that constant in case of the nucleons. If the space displacement is constrained, then we obtain the angular velocity

$$\mathbf{w} = -\sin\frac{d}{R}\cdot\frac{v_0R}{2d^2}\frac{\mathbf{d}}{d}. \tag{12}$$

The angular velocity (12) is the reason that the ecliptic plane of planetary orbits is rotated with respect to the galactic plane. The acceleration (11) has important role for the motion of the stars in the galaxies such that we do not need to introduce dark matter and to bind the stars on a distance at most πR, where the acceleration is attractive [16].

It is interesting that R. Boscovich in his Ref. [1] considered also four basic cases between the space and time which are related to one point and analogous to them also four cases which are related for several points. He also comments which combinations, i.e. compositions among these cases are possible and which are not possible. There is an interesting analogy between his comments and the results in this Chapter based on the Lie groups and their properties.

4. Magnetic fields of the spinning bodies

First we give the following preliminaries about the magnetic fields. While in case of gravitation and inertial forces we have motion in the space $S \times T$, the electromagnetic interaction occurs in the space $SR \times T$ where the antisymmetric matrix F_{ij} consisting of electric and magnetic field behaves as a tensor in 3 + 1 = 4 dimensions or alternatively in 3 + 3 = 6 dimensions. But however, we observe motion and angular velocity of the charged particles in the observable $S \times T$ space. If $\mathbf{E} = 0$, according to Ref. [18] or (38), the observed angular velocity is equal to $\Omega = \frac{e}{mc}\mathbf{H}$, where m and e are the mass and the charge of the electron or the considered charged body. This formula assumes that the density of mass is proportional to the density of charge at each point. This is satisfied in case of the electron. But in case of proton and neutron these two densities are no globally proportional, and so there appear the factors $g_p \approx 2.792$ and $g_n \approx -1.913$ in front of $\frac{e}{mc}\mathbf{H}$.

Further let us consider again the case $r \to t$. If the angular velocity of the electron may take arbitrary value Ω, i.e. it is not constrained, then an electron in magnetic field would move with a constant velocity ($\mathbf{v}$) and should rotate with the angular velocity $\Omega = \frac{e}{mc}\mathbf{H}$. The force $\frac{e}{c}\mathbf{v} \times \mathbf{H}$ as a part of the Lorentz force would disappear. Assuming that the unadmitted angular velocity is $\Omega = \frac{e}{mc}\mathbf{H}$ for $\mathbf{E} = 0$, then according to the acceleration $\mathbf{v} \times \mathbf{w}$, i.e. "temporal part" of the Coriolis acceleration, we obtain that the induced displacement is caused by the force

$$\mathbf{f} = m(-\Omega) \times \mathbf{v} = m\frac{-e}{mc}\mathbf{H} \times \mathbf{v} = \frac{e}{c}\mathbf{v} \times \mathbf{H}. \tag{13}$$

So, as a consequence of the case $r \to t$ we obtain the Lorentz force in magnetic field. Now the general formula for the Lorentz force in case of arbitrary electromagnetic field is a consequence of the Lorentz covariance of the Lorentz force among the inertial coordinate systems.

Indeed the magnetic field $\mathbf{H}$ manages partially to rotate the charged particles. The charge e may be defined as a coefficient of proportionality between the unadmitted angular velocity Ω and $\mathbf{H}/(mc)$, which are collinear.

Finally we can resume the following. A charged body in a magnetic field $\mathbf{H}$ should rotate with angular velocity $\Omega = \frac{e}{mc}\mathbf{H}$. On the other hand, the Lorentz force is a consequence of the argument that the unadmitted angular velocity of the charged body is the reason of the existence of the Lorentz force. If we assume that the unadmitted angular velocity is also $\Omega = \frac{e}{mc}\mathbf{H}$, we obtain in (13) a satisfactory result. Hence we come to the conclusion that both admitted and unadmitted angular velocity are equal to $\Omega = \frac{e}{mc}\mathbf{H}$. So the total angular velocity (admitted and unadmitted) is equal to

$$\Omega_t = \frac{2e}{mc}\mathbf{H}. \tag{14}$$

In order to study the magnetic field of the spinning bodies, it is necessary to refer to the precessions of gyroscopes. In Refs. [18–20] the precessions in case of Gravity Probe B (GPB) experiment were considered. The total precession (geodetic precession and frame dragging) of the gyroscope with respect to the chosen coordinate system is deduced to be

$$\Omega_{gyr} = 2\sum_a (\mathbf{v} - \mathbf{v}_a) \times \nabla \frac{Gm_a}{r_a c^2} - \sum_a G[\mathbf{J}_a - 3\hat{\mathbf{n}}_a(\hat{\mathbf{n}}_a \cdot \mathbf{J}_a)]/r_a^3 c^2, \tag{15}$$

where $a = 1,2,3,\cdots$ denotes the a-th celestial spherical and homogeneous body which causes this precession, $\mathbf{v}$ denotes the velocity of the observer, $\mathbf{v}_a$ denotes velocity of the barycentre of the a-th celestial body, $\mathbf{J}_a$ is its angular momentum, m_a is its mass, r_a is the distance from it to the observer, and $\hat{\mathbf{n}}_a$ is the unit radial vector to the gyroscope. In non-integrated form this angular velocity is given by

$$\Omega_{gyr} = 2\sum_i \frac{(\mathbf{v} - \mathbf{u}_i) \times \mathbf{a}_i}{c^2}, \tag{16}$$

where $\mathbf{a}_i$ is the acceleration toward the i-the particle and $\mathbf{u}_i$ is the velocity of the i-th particle. The coefficient "2" is analogous to the coefficient "2" in case of Coriolis force, i.e. half of this value corresponds to the changes in the space and other half corresponds to the changes in time. According to Ref. [20] it also appears a precession of the local coordinate system close to the massive bodies, which in non-integrated form is given by

$$\Omega_{lcs} = \frac{1}{2}\sum_i \frac{(\mathbf{v} - \mathbf{u}_i) \times \mathbf{a}_i}{c^2}. \tag{17}$$

It is analogous to the Thomas precession. The precession (17) leads to apparent precession of the celestial bodies on the sky. It is necessary to consider this precession, because otherwise the total precession (geodetic plus frame dragging) would not be Lorentz covariant. This precession is unknown for the General Relativity. C.M. Will in his book ([21], p. 95) noticed that the term for the frame dragging deduced in General Relativity is not Lorentz covariant, but he did not offer any solution of the problem.

Analogously to (17), in integrated form this precession is given by

$$\Omega_{lcs} = \frac{1}{2}\sum_a (\mathbf{v} - \mathbf{v}_a) \times \nabla \frac{Gm_a}{r_a c^2} - \frac{1}{4}\sum_a G[\mathbf{J}_a - 3\hat{\mathbf{n}}_a(\hat{\mathbf{n}}_a \cdot \mathbf{J}_a)]/r_a^3 c^2. \tag{18}$$

The precessions (15) and (18) have basic theoretical role, but in case of experimental measurements, as GPB experiment, it is measured the relative precession of the gyroscope (15) with respect to the local coordinate system (18), i.e. it is measured their subtraction

$$\Omega_{\text{rel}} = \frac{3}{2}\sum_{\text{a}}(\mathbf{v} - \mathbf{v}_{\text{a}}) \times \nabla\frac{Gm_{\text{a}}}{r_{\text{a}}c^2} - \frac{3}{4}\sum_{\text{a}} G[\mathbf{J}_{\text{a}} - 3\hat{\mathbf{n}}_{\text{a}}(\hat{\mathbf{n}}_{\text{a}}\cdot\mathbf{J}_{\text{a}})]/r_{\text{a}}^3c^2. \tag{19}$$

So the geodetic precession now remains the same as in case of the General Relativity, while the precession of the frame dragging is 25% less than the value from the General Relativity.

The relative precession (19) is confirmed by the GPB experiment according to the following comments.

	geodet. Prec. (mas/yr)	**frame drag. (mas/yr)**
gyroscope 1	-6588.6 ± 31.7	-41.3 ± 24.6
gyroscope 2	-6707.0 ± 64.1	-16.1 ± 29.7
gyroscope 3	-6610.5 ± 43.2	-25.0 ± 12.1
gyroscope 4	-6558.7 ± 33.2	-49.3 ± 11.4
predict. From GR	-6606.1	-39.2
predict. From (19)	-6606.1	-29.4

The final results from the measurements of the geodetic precession and the frame dragging are given by the previous table published in [22]. In the last row now for comparison it is added the prediction according to Eq. (19). There is no doubt about the formula for the geodetic precession, which is -6606.1 mas/yr. The third gyroscope gives the most close result to this value. It means that it has the best of the required performances, for example it should be homogeneous and spherical body. So it is naturally to expect that this gyroscope gives the most precise results about the frame dragging. The measured value for the frame dragging is (-25.0 ± 12.1) mas/yr and it is closer to the prediction of -29.3 mas/yr according to Eq. (19), than the prediction from the General Relativity, which is -39.2 mas/yr. So, the GPB experiment confirms the precession (18), which will be essential in deducing the magnetic field of the spinning bodies.

Now we are ready to deduce the magnetic field of the spinning bodies. It is analogous to the deduction of the spin velocities. Let us consider separately local coordinate systems for all particles of the Earth, or arbitrary spinning celestial body. We use the previous formulas where the celestial bodies are the atoms of the spinning body (Earth) as micro celestial bodies. Using that the particles of the a-th celestial body move with average velocity $\mathbf{v}_{\text{a}}$ of the barycentre of that celestial body, and using that $\mathbf{J}_{\text{a}} \approx 0$ for each such small particle, i.e. its self angular momentum is almost 0, according to (18) we obtain that each local coordinate system intends to rotate with average angular velocity $\frac{1}{2}\sum_{\text{a}}(\mathbf{v} - \mathbf{v}_{\text{a}}) \times \nabla\frac{Gm_{\text{a}}}{r_{\text{a}}c^2}$. But globally, all local coordinate systems rotate with the same angular velocity (18), because the Earth is a solid celestial body. So for each local coordinate system the non-inertial constrained angular velocity is $-\frac{1}{4}\sum_{\text{a}} G[\mathbf{J}_{\text{a}} - 3\hat{\mathbf{n}}_{\text{a}}(\hat{\mathbf{n}}_{\text{a}}\cdot\mathbf{J}_{\text{a}})]/r_{\text{a}}^3c^2$. Half of the corresponding inertial angular velocity, which is with opposite sign, will be admitted although non-observable, and the other half of the inertial angular velocity from the spatial part is non-admitted and it induces magnetic field. Hence the magnetic field of the Earth is a sum of large number of such small magnetic fields caused by each particle of the Earth. In order to determine the magnetic field, the angular velocity $\frac{1}{4}\sum_{\text{a}} G[\mathbf{J}_{\text{a}} - 3\hat{\mathbf{n}}_{\text{a}}(\hat{\mathbf{n}}_{\text{a}}\cdot\mathbf{J}_{\text{a}})]/r_{\text{a}}^3c^2$ should be multiplied by $m/(2e)$ according to (14). Here m is the mass of each particle of the Earth and e is the charge which "corresponds" to m.

The quotient m/e we chose to be the universal constant $1/\sqrt{4\pi\varepsilon_0 G}$, such that $Gm^2/R^2 = e^2/(4\pi\varepsilon_0 R^2)$, i.e. the Newton force is equal to the Coulomb force up to the sign. Hence the magnetic field measured from arbitrary inertial coordinate system is given by $\mathbf{H} = c\mathbf{B}$, where

$$\mathbf{B} = \frac{-G}{8c^2 r^3 \sqrt{4\pi\varepsilon_0 G}} [\mathbf{J} - 3\hat{\mathbf{n}}(\hat{\mathbf{n}} \cdot \mathbf{J})]. \tag{20}$$

Let the spinning axis is chosen as z-axis and if we put $\hat{\mathbf{n}} = (\cos\varphi,\ 0,\ \sin\varphi)$, then

$$\mathbf{B} = \frac{-J\sqrt{G}}{8c^2 r^3 \sqrt{4\pi\varepsilon_0}} \left(-3\cos\varphi\sin\varphi, 0, 1 - 3\sin^2\varphi\right). \tag{21}$$

If r is a constant, for example r is radius of the Earth, then

$$B = |\mathbf{B}| = \frac{J\sqrt{G}}{8c^2 r^3 \sqrt{4\pi\varepsilon_0}} \sqrt{1 + 3\sin^2\varphi}. \tag{22}$$

The calculations lead to the following values: 0.295G on the equator ($\varphi = 0$) and 0.59G on the poles ($\varphi = \pi/2$), which fits with the measured values.

This magnetic field is the same in arbitrary inertial coordinate system. So, this magnetic field may not be a part of electromagnetic tensor field. It means that in a moving coordinate system with arbitrary velocity close to the Earth an electric field **E** will not appear, and electromotive force can not be obtained.

The previous conclusion is deduced if the spinning body (Earth) is electro-neutral body. If there are some free electrons on the Earth and inside, since they are rotating together with the Earth they will produce a small quantity of magnetic field, and this magnetic field may be part of tensor of electromagnetic field, because it is caused by charged particles. The telluric current is not caused by the geomagnetic field, but from the electrostatic properties of the Earth, atmosphere and solar radiation.

In general case, as we will see in the next section, the electromagnetic field is a consequence of 3*. Hence the magnetic field of spinning bodies belongs to the case 3*. It is a specific case because the produced magnetic field can not be a part of the tensor of the electromagnetic field and consequently there is no magnetic induction.

At the end of this section let us return to the admitted half of the inertial angular velocity $\frac{1}{4}\sum_a G[\mathbf{J}_a - 3\hat{\mathbf{n}}_a(\hat{\mathbf{n}}_a \cdot \mathbf{J}_a)]/r_a^3 c^2$. Although this admitted half is constrained in case of the Earth, it is not constrained for the gyroscope. So, the coefficient -1 from (15) becomes now $-1 + \frac{1}{8} = -\frac{7}{8}$, and hence the coefficient $-\frac{3}{4}$ from (19) becomes $-\frac{5}{8}$. The new prediction for the frame dragging is now $-39.2 \times \frac{5}{8} = -24.5$ mas/yr, which almost the same with the average measured value -25 mas/yr for the third gyroscope.

5. Electromagnetic fields of the charged bodies

Let us return to the Eq. (7) with the same notations. Then the first charged body rotates the second charged body. We assume as an Axiom 1 that the angle of rotation is equal to (**Figure 4**)

$$\vec{\theta} = \frac{e_1 e_2}{4\pi\varepsilon_0 r^2 m_2 c^2} \mathbf{r}, \tag{23}$$

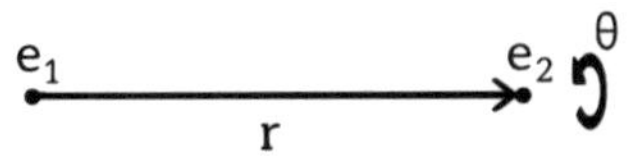

Figure 4.
The second charged particle is rotated by the first charged particle (and vice versa).

where $\mathbf{r} = (x, y, z)$ is the radius vector from the center of the first charged body toward the second body. This axiom will be used for deriving the Coulomb's force. So, this geometrical assumption gives us a more fundamental understanding of the charge as a physical phenomena.

Let us consider two charged particles with centers at O_1 and O_2 and let X be a close point of O_2 of the second particle with radius vector $(x + a, y + b, z + c)$, where a, b, and c are small quantities of length. We can use again the **Figure 3**, although it was related for strong interaction. Then it causes displacement of the point X by the vector $\theta\frac{(x,\ y,\ z)}{r} \times (x + a, y + b, z + c) = \theta\frac{(x,\ y,\ z)}{r} \times (a, b, c)$, and it corresponds to angle $\theta\frac{(x,\ y,\ z)}{r^2} \times (a, b, c)$.

This is a consequence since the matrices of Lorentz boosts in general do not commute with the matrices of rotation from Section 1. Moreover, we can use Galilean transformations instead of Lorentz transformations and choose the coordinate system such that the angle of rotation is $(0, 0, \theta)$ and obtain

$$\begin{bmatrix} \cos\theta & -\sin\theta & 0 & 0 \\ \sin\theta & \cos\theta & 0 & 0 \\ 0 & 0 & 1 & 0 \\ 0 & 0 & 0 & 1 \end{bmatrix} \begin{bmatrix} 1 & 0 & 0 & a \\ 0 & 1 & 0 & b \\ 0 & 0 & 1 & c \\ 0 & 0 & 0 & 1 \end{bmatrix}$$

$$\begin{bmatrix} \cos\theta & -\sin\theta & 0 & 0 \\ \sin\theta & \cos\theta & 0 & 0 \\ 0 & 0 & 1 & 0 \\ 0 & 0 & 0 & 1 \end{bmatrix}^{-1} \begin{bmatrix} 1 & 0 & 0 & a \\ 0 & 1 & 0 & b \\ 0 & 0 & 1 & c \\ 0 & 0 & 0 & 1 \end{bmatrix}^{-1} \approx \begin{bmatrix} 1 & 0 & 0 & a' \\ 0 & 1 & 0 & b' \\ 0 & 0 & 1 & c' \\ 0 & 0 & 0 & 1 \end{bmatrix},$$

where $(a', b', c') = (0, 0, \theta) \times (a, b, c)$.

The unadmitted translation leads to the Coulomb's acceleration/force

$$\mathbf{a} = \frac{c^2}{2}\mathrm{rot}\left(\theta\frac{(x,\ y,\ z)}{r^2} \times (a,\ b,\ c)\right) = \theta c^2\frac{(x, y, z)}{r^2} = \frac{e_1 e_2}{4\pi\varepsilon_0 r^3 m_2}(x, y, z), \tag{24}$$

$$\mathbf{f} = \frac{e_1 e_2}{4\pi\varepsilon_0 r^3}(x, y, z). \tag{25}$$

Hence the electric field caused by the first charged body is

$$\mathbf{E} = \frac{e_1}{4\pi\varepsilon_0 r^3}(x, y, z). \tag{26}$$

Since the angle of rotation observed orthogonally to the plane of the angle, remains unchanged, the charge remains unchanged in all inertial coordinate systems.

The special case when $e_2/m_2 = \sqrt{4\pi\varepsilon_0 G}$ is of special interest. Then (23) leads to

$$\theta_* = \frac{e\sqrt{G}}{\sqrt{4\pi\varepsilon_0} rc^2}, \tag{27}$$

and the corresponding acceleration is given by

$$\mathbf{a}_* = \frac{e\sqrt{G}}{\sqrt{4\pi\varepsilon_0} r^3}(x, y, z). \tag{28}$$

This is extremely small acceleration. If we compare the acceleration of the electron toward arbitrary charged particle, the acceleration (28) is smaller for coefficient

$$\frac{m_e}{e} \frac{\sqrt{G}}{(4\pi\varepsilon_0)^{3/2}} = 4.8994 \cdot 10^{-22}. \tag{29}$$

We introduce now Axiom 2, which will be used for deriving the Biot-Savart law. This axiom states that any elementary particle (not composite) with mass m has charge e such that

$$|e| \geq m\sqrt{4\pi\varepsilon_0 G}, \tag{30}$$

i.e. $\sqrt{4\pi\varepsilon_0 G}$ is the minimal possible value for the quotient e/m. This axiom does not refer to composite particles, and also arbitrary electro-neutral particle should contain both electro-positive part and electro-negative part. This constraint is analogous to the constraint $|v| \leq c$. The local coordinate system, considered as local frame, can be treated as a border case: $m = 0$, $e = 0$ and $e/m = \sqrt{4\pi\varepsilon_0 G}$. So, (27) and (28) can be written as $\theta_{\text{lcs}} = \frac{e\sqrt{G}}{\sqrt{4\pi\varepsilon_0} rc^2}$ and $\mathbf{a}_{\text{lcs}} = \frac{e\sqrt{G}}{\sqrt{4\pi\varepsilon_0} r^3}(x, y, z)$. The Axiom 2 is in accordance with the previous section, where we used the universal constant $1/\sqrt{4\pi\varepsilon_0 G}$ for the quotient m/e.

If the charged particle moves with velocity $\mathbf{u}$, then we have angular velocity

$$\mathbf{w}_{\text{lcs}} = \mathbf{u} \times \frac{e\sqrt{G}}{\sqrt{4\pi\varepsilon_0} r^3 c^2}(x, y, z), \tag{31}$$

which is rather different than (17). Here it is used that the acceleration together with the components of the angular velocity multiplied by c form a tensor ϕ_{ij}, which is analogous to the tensor of the electromagnetic field F_{ij}. Indeed it is an antisymmetric tensor since it is the Lie algebra of G_t as 4×4 matrix group with imaginary temporal coordinate. Analogously to the deduction of the magnetic field of the spinning bodies, now we should multiply this angular velocity by the universal constant $m/e = 1/\sqrt{4\pi\varepsilon_0 G}$, and obtain the Biot-Savart law (33). Moreover, since ϕ_{ij} is a tensor, $-\mathbf{v} \times \mathbf{w}$ is acceleration, which in electromagnetism means that $(-\mathbf{v}) \times \frac{\mathbf{H}}{c}$ is electric field. So, for a charged body in electric field $\mathbf{E}$ and magnetic field $\mathbf{H}$, in local system where the charged body rests, appears also an electric field $\frac{\mathbf{v}}{c} \times \mathbf{H}$ and we obtain the Lorentz force

$$\mathbf{f} = e\left(\mathbf{E} + \frac{\mathbf{v}}{c} \times \mathbf{H}\right). \tag{32}$$

This is the case 3*, because the non-compatibilities arising from the space rotations in the temporal group $SR \times T$ (G_t) lead to some changes in the same space $SR \times T$, which is observed as Coulomb's force and Lorentz force. The electromagnetic tensor field F_{ij} in the Lie group $SR \times T$, plays the main role among the inertial coordinate systems.

At the end of this section we give also a deeper view of the electromagnetic field by returning to the gyroscopes. Let us consider a charged particle with mass m and charge e and let us find the magnetic field caused by the angular velocity (16). Half of it is admitted and the other half is non-admitted and it causes the required magnetic field. It means constraint of both geodetic precession and frame dragging. While the coefficient m/e was universal constant when it was applied to (18), in this case m/e is determined by the mass and charge of the considered charged particle. The magnetic field appears, because for the charged particles, it is not admitted to rotate completely. In order to simplify the calculation, we consider only one body ($a = 1$) with point mass m, charge e and hence $J = 0$. First let us consider the border case $e/m = \frac{1}{\sqrt{4\pi\varepsilon_0 G}}$. Analogously to the previous section, first we multiply the Eq. (16) by $\frac{1}{2} \frac{1}{\sqrt{4\pi\varepsilon_0 G}}$ and replace $\mathbf{v} = 0$ because we measure the magnetic field with respect to us. Further in general case, for arbitrary mass and charge, using the proportionality $B \sim e$, the Eq. (16) should additionally be multiplied by $\frac{e}{m} \frac{1}{\sqrt{4\pi\varepsilon_0 G}}$. It means that in the last two steps the Eq. (16) is multiplied by $\frac{1}{2} \frac{e}{m} \frac{1}{4\pi\varepsilon_0 G}$, and hence it becomes

$$\mathbf{B} = \frac{\mathbf{H}}{c} = \mathbf{u} \times \nabla \frac{e}{4\pi\varepsilon_0 r c^2}, \tag{33}$$

which is the Biot-Savart law. So the Biot-Savart law is a consequence from constraint in angular velocity of the body with electric charge e. If we have electric and magnetic fields $\mathbf{E}$ and $\mathbf{H}$, having in mind the formula (33) and $\Omega = \frac{e}{mc}\mathbf{H}$ (when $\mathbf{E} = 0$), its generalization leads to

$$\Omega = \frac{e}{mc}\left(\mathbf{H} - \frac{1}{c}(\mathbf{v} \times \mathbf{E})\right). \tag{34}$$

Note that while the magnetic field of the spinning bodies was obtained from Eq. (17), the Biot-Savart law was obtained from Eq. (16). Analogously as in case of GPB experiment where the subtraction $\Omega_{\text{gyr}} - \Omega_{\text{lcs}}$ was measured, here the measurable magnetic induction can be obtained from the subtraction of the magnetic inductions (20) and (33).

6. Gravitation and general equations of motions

Although the gravitation is well studied phenomena, now we explain the mechanism of the attraction. The gravitation field caused by a mass M induces that on distance r from the center the lengths seem to be larger for the coefficient $1 + \gamma \frac{GM}{rc^2}$, if the terms of order c^{-4} and smaller are neglected. Here γ is the known PPN parameter and takes value 1.

Let us consider a body with mass M and center at point O, and let us consider also a test body with negligible mass and center at A and radius vector $\overrightarrow{OA} = (x, y, z)$. Let B be a close point to A of the test body with radius vector $\overrightarrow{OB} = (x + a, y + b, z + c)$, where a, b, and c are small quantities of length (**Figure 5**). The vector product of almost unit vectors

$$\frac{(x, y, z)}{r} \quad \text{and} \quad \frac{(x+a, y+b, z+c)}{r},$$

where $r = \sqrt{x^2 + y^2 + z^2}$, leads approximately to the angle

$$\angle AOB = \frac{(x, y, z)}{r} \times \frac{(x+a, y+b, z+c)}{r}. \tag{35}$$

The vector (a, b, c), far from the massive bodies is observed as

$$\left(a\left(1+\gamma\frac{GM}{rc^2}\right), b\left(1+\gamma\frac{GM}{rc^2}\right), c\left(1+\gamma\frac{GM}{rc^2}\right)\right)$$

and instead of (35) we have the angle

$$\angle AOB' \approx \frac{(x, y, z)}{r} \times \frac{\left(x+a\left(1+\frac{\gamma GM}{rc^2}\right), y+b\left(1+\frac{\gamma GM}{rc^2}\right), z+c\left(1+\frac{\gamma GM}{rc^2}\right)\right)}{r}. \tag{36}$$

The subtraction between the vectors (35) and (36) is given by

$$\angle BOB' = \gamma\frac{GM}{rc^2}\frac{(yc - zb, za - xc, xb - ya)}{r^2}.$$

Half of this angle is admitted and the other half is not admitted. The unadmitted half induces acceleration given by

$$\mathbf{g} = \gamma\frac{c^2}{2}\mathrm{rot}\angle BOB' = -\gamma c^2\frac{GM}{rc^2}\frac{(x, y, z)}{r^2},$$

i.e.

$$\mathbf{g} = -\gamma\mathbf{r}\frac{GM}{r^3}. \tag{37}$$

Now we are able to explain why the two different deductions of the Biot-Savart law from the previous section led to identical result. It is sufficient to consider only the special case when $e/m = \sqrt{4\pi\varepsilon_0 G}$ and to prove that the gravitational acceleration **a**

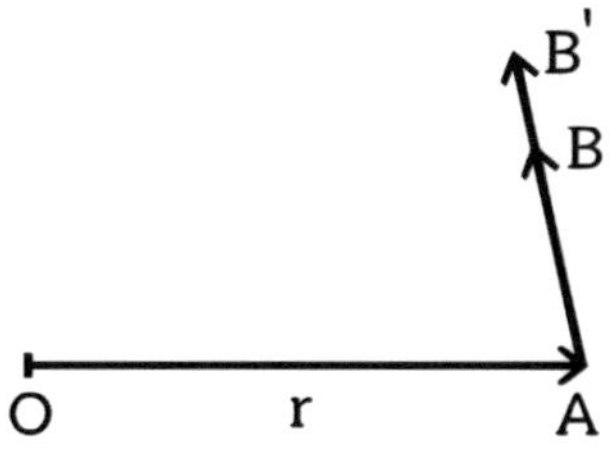

Figure 5.
Observation of lengths in gravitational field.

from (16) is identical with the Coulomb acceleration. So, we need to prove that the gravitational acceleration $\mathbf{a}_{grav} = \frac{c^2}{2}\mathrm{rot}\angle BOB'$ from (37) is identical with the Coulomb's acceleration $\mathbf{a}_{el} = \frac{c^2}{2}\mathrm{rot}\left(\theta\frac{(x,\ y,\ z)}{r^2} \times (a,\ b,\ c)\right)$ from (24), neglecting the terms of order c^{-4}. Indeed it is easy to verify that $\angle BOB' = \theta\frac{(x,\ y,\ z)}{r^2} \times (a, b, c)$ neglecting the terms of order c^{-4}. This argument makes the Axiom 1 more evident.

Shortly we discuss what happens with the admitted half of the angle $\angle BOB'$. It leads to the total precession of the gyroscopes known as geodetic precession and frame dragging. It is analogous to the magnetic field in case of the electromagnetism. For example the geodetic precession is given by $\Omega = -\frac{3}{2}(\mathbf{v} \times \mathbf{a})/c^2$. The coefficient 3/2 now is not the same with 1 in case of the magnetitic field, because the coefficient 3/2 includes the PPN parameter $\gamma = 1$, while in case of the electromagnetism $\gamma = 0$, because the rotation caused by the charged bodies does not changes the Euclidean metric $g_{11} = g_{22} = g_{33} = 1$.

There are two different approaches for the equations of motion: (i) The motion of the particles is according to the metric of the massive body and (ii) the metric caused by the massive body should not be used for motion of the particles. The first approach is widely used in the General Relativity. The second approach is proposed by Logunov [23]. It is also used in [24] and there for equations of motion is used a spin connection in the space–time. It is based on the antisymmetric tensor ϕ_{ij} with respect to the Lorentz transformations, which can be used now. We mentioned that it is analogous to the tensor of electromagnetic field, where instead of electric field we have acceleration and instead of magnetic field we have angular velocity multiplied by c. So, this approach is analogous to the electromagnetism, and this connection can be used also for rotating systems. Moreover, in [18] the equations of motion are converted into high-dimensional space–time. Using the gravitational acceleration and the Lorentz covariance, the precession (16) occurs.

The change of the metric for the time in case of electromagnetism is obvious from (7), but in case of electromagnetism we do not use the metric for motion of the charged bodies. On the other hand the experience shows that the metric in gravitation can be used for equations of motion. It occurs because the force in gravitation is proportional to the mass, but not with the charge as in electromagnetism.

This is the case 4*, because the non-compatibilities arising from the lengths in the temporal group $S \times T$, i.e. G_t, lead to some changes in the same space $S \times T$, which is observed as gravitational force and precessions (geodetic precession and frame dragging). The gravitational force is very close to the inertial forces, and so in the General Relativity the Equivalence Principle was introduced.

Having in mind the previous comments we present now the equations of motions and precessions in case of gravitation, and on the other hand they can be used for motions and precessions of the charged particles in electromagnetic field. It refers to test bodies with negligible mass. The following equations of motion are done with respect to the three-dimensional time and they are in agreement of the gravitational experiments known from the General Relativity. The general 3×3 complex matrix equation for motion and precession of test particles is given by

$$\frac{\mathrm{d}(\exp(iA + C) - \exp(-iA - C))}{\mathrm{d}s} = \frac{1}{\sqrt{1 - v^2/c^2}} S + \frac{\mathbf{v}}{c\sqrt{1 - v^2/c^2}} \times S, \tag{38}$$

where A is antisymmetric 3×3 matrix which corresponds to motion with velocity, i.e.

$$e^{iA} = \frac{1}{\sqrt{1-\frac{v^2}{c^2}}}\begin{bmatrix} 1 & 0 & 0 \\ 0 & 1 & 0 \\ 0 & 0 & 1 \end{bmatrix} - \frac{1}{(1+\gamma)(c^2-v^2)}\begin{bmatrix} v_x^2 & v_xv_y & v_xv_z \\ v_xv_y & v_y^2 & v_yv_z \\ v_xv_z & v_yv_z & v_z^2 \end{bmatrix} - \frac{i}{c\sqrt{1-\frac{v^2}{c^2}}}\begin{bmatrix} 0 & v_z & -v_y \\ -v_z & 0 & v_x \\ v_y & -v_x & 0 \end{bmatrix},$$

C is antisymmetric 3×3 matrix which corresponds to space rotation, i.e. $\Omega = d\mathbf{C}/dt$, $ds = cdt\sqrt{1 - v^2/c^2}$, and $\mathbf{v}$ is 3-vector of velocity. The matrix S is defined by $S = Q^T\phi Q$, where the tensor ϕ as a complex 3×3 matrix is given by

$$\phi = \begin{bmatrix} 0 & \frac{w_z}{c} - i\frac{a_z}{c^2} & -\frac{w_y}{c} + i\frac{a_y}{c^2} \\ -\frac{w_z}{c} + i\frac{a_z}{c^2} & 0 & \frac{w_x}{c} - i\frac{a_x}{c^2} \\ \frac{w_y}{c} - i\frac{a_y}{c^2} & -\frac{w_x}{c} + i\frac{a_x}{c^2} & 0 \end{bmatrix}. \tag{39}$$

The matrix Q depends on the relative velocity of the particle with respect to the source of the field, and becomes I if the relative velocity is 0. It is determined in [18]. The matrix $\mathbf{v} \times S$, for each antisymmetric matrix S is defined by $f^{-1}(\mathbf{v} \times (s_{23}, s_{31}, s_{12}))$, where $f(L) = (L_{23}, L_{31}, L_{12})$ for each antisymmetric matrix L.

The PPN parameter γ has the same role in case of equations of motion (35) as the matrices Q and Q^T. Indeed if $\gamma = 1$, then the matrices Q and Q^T should be included (in case of gravitation) and if $\gamma = 0$ (in case of electromagnetism), then Q and Q^T should be omitted. This is noticed in [24], where instead of 3×3 matrices Q is considered its equivalent 4×4 matrix P. In case of electromagnetism the matrix S should be the following

$$S = -\frac{e}{mc^2}\begin{bmatrix} 0 & E_z + iH_z & -E_y - iH_y \\ -E_z - iH_z & 0 & E_x + iH_x \\ E_y + iH_y & -E_x - iH_x & 0 \end{bmatrix}. \tag{40}$$

Then the Lorentz force (32) and the angular velocity of the charged body (34) can be obtained.

7. Conclusions

This research gives a global view of the interactions in the nature, especially the electromagnetic. It is based of the non-commutativity of the structural Lie groups G_t and G_s. The duality is one of the main principle of this model. If the non-inertial velocity (or angular velocity) is constrained, then it transforms into inertial velocity (angular velocity), which belongs to the temporal part of the space–time. Also the admitted and non-admitted quantities are equal. Indeed, both these quantities are mutually dependent. It makes the model more abstract, but it leads to satisfactory results and gives more fundamental view of the interactions. It also gives new view of the charges, such that a charged particle rotates the other charged particles for a small angle θ. The magnetic field of the spinning bodies is separated from the electromagnetic field of the charged particles. It also explains why the charge remains the same among the inertial systems. This view helps us more clearly to see the similarity between the electromagnetism and gravitation and also to see where they differ. In some future research should be considered the electromagnetic waves from this geometrical viewpoint.

Author details

Kostadin Trenčevski
Ss. Cyril and Methodius University of Skopje, Skopje, Macedonia

*Address all correspondence to: kostadin.trencevski@gmail.com

References

[1] Bošković R. O prostoru vremenu i relativnosti. Beograd: Kultura; 1956. (original: De spatio, ac tempore. Ut a nobis cognoscuntur. De motu absoluto, an possit a relativo distingui. De vi inertiae.)

[2] Trenčevski K. Duality in the special relativity based on the isomorphic structural groups $SO(3, \mathbb{C})$ and $O_+^{\uparrow}(1, 3)$. Tensor. 2010;**72**(1):32-46

[3] Trenčevski K. Special relativity based on the $SO(3, C)$ structural group and 3-dimensional time. Mathematica Balkanica. 2011;**25**(1-2):193-201

[4] Trenčevski K. Representation of the Lorentz transformations in 6-dimensional space-time. Kragujevac Journal of Mathematics. 2011;**35**(2): 327-340

[5] Trenčevski K. On the geometry of the space-time and motion of the spinning bodies. Central European Journal of Physics. 2013;**11**(3):296-316

[6] Trenčevski K. Application of the geometry of curves in Euclidean space. Filomat. 2019;**33**(4):1029-1036. DOI: 10.2298/FIL1904

[7] Trenčevski K. On the group of isometries of the space. In: Proceedings of the Conference Differential Geometry - Dynamical Systems (DGDS2013); 10-13 October 2013; Bucharest: BSG Conf. Proc. 21; 2014. p. 193-200

[8] Yefremov A. Bi-quaternion square roots, rotational relativity, and dual space-time intervals. Gravitation and Cosmology. 2007;**13**(3):178-184

[9] Yefremov A. Notes on pioneer anomaly explanation by satellite shift formula of quantum relativity: Remarks on less mundane explanation on pioneer anomaly from Q-relativity. Progress in Physics. 2007;**2**:93

[10] Barashenkov VS. Quantum field theory with three-dimensional vector time. Particles and Nuclei, Letters. 2004; **2**:54-63

[11] Cheng X. Three dimensional time theory: To unify the principles of basic quantum physics and relativity. 2005: arXiv: quant-ph/0510010

[12] Cheng X. Modified Kaluza-Klein theory, quantum hidden variables and 3-dimensional time. 2005: arXiv: quant-ph/0501034

[13] Kitada H, Fletcher LR. Unification of physics part I. Local time. Apeiron. 1996; **3**(2):38-45. arxiv: gr-qc/0110065

[14] Kitada H. Three dimensional time and energy operators and uncertainty relation. 2000; arXiv: quant-ph/0007028

[15] Tegmark M. On the dimensionality of the spacetime. Classical and Quantum Gravity. 1997;**17**:L69-L75

[16] Trenčevski K. Global scheme of the basic interactions and their geometrical interpretations. presented on XXI Geometrical Seminar, 26.6-2.7.2022 Belgrade, submitted for publ.

[17] Trenčevski K, Celakoska E. Induced spin velocity of the Earth and its influence to the seasonal variation of the Earth's angular velocity. The European Physical Journal Plus. 2020;**135**:450

[18] Trenčevski K, Celakoska E. Complex equations of motion for a body under gravitational influence by using nine-parameter space-time bundle with structure group $SO(3, \mathbb{C})$. Annals of Physics. 2018;**395**:15-25

[19] Trenčevski K. Magnetic fields of the spinning bodies. International Journal of Geometric Methods in Modern Physics. 2015;**12**(4):1550046. 9 p

[20] Trenčevski K, Celakoska E. Geodetic precession and frame dragging observed far from massive objects and close to a gyroscope. Central European Journal of Physics. 2011;**9**:654-661

[21] Will CM. Theory and Experiment in Gravitational Physics. New York: Cambridge University Press; 1993

[22] Everitt CWF et al. Gravity probe B: Final results of a space experiment to test general relativity. Physical Review Letters. 2011:106, 221101 arxiv:1105.3456

[23] Logunov AA, Mestvirishvili MA. Relativistic Theory of Gravity. Moscow: Nauka; 1989

[24] Trenčevski K, Celakoska E, Balan V. Research of gravitation in flat Minkowski space. International Journal of Theoretical Physics. 2011;**50**(1):1-26

Chapter 2

Proper Understanding of the Natures of Electric Charges and Magnetic Flux

Salama AbdelHady

Abstract

According to an analogy between the laws that characterize the flow of heat, electric charges, and magnetic flux, and to results of Faraday's experiments, the electric charge and the magnetic flux were defined and visualized in previous research as energy flux in the form of electromagnetic waves (EM) that have electric or magnetic potentials and that all potentials can be measured by the Volt. The proofs of such statements will be enlightened in this chapter. Recognizing the nature of electric charges as energy; the electric current, defined traditionally as the rate of flow of electric charges would have the unit of power, i.e., Watt. As the ammeter does not measure the power but measures, according to its definition, the quotient of the electric power divided by the electric potential, then its unit should be "Watt/Volt." So, the ammeter does not measure an electric current if it is defined as the rate of flow of electric charges. However, the unit "Watt/Volt" of the ammeter's readings is distinguished as a unit of a property of the electric field that defines the capacity of the electric field to allow flow of a definite rate of electric energy by force of a unit of the electric potential, i.e., by 1 Volt. It will be shown in the presented study that this capacity measures also the rate of growth of one of the physical properties of the electric field which is called "entropy." Hence, the ammeter measures acceptability of the electric field to the flow of electric power and measures the rate of generation of entropy, or destruction of exergy of the measured electric field. By analogy between the electric and magnetic energies, it will be proved that the magnetic fluxmeter measures also the quotient of the magnetic power divided by the magnetic potential and its unit "Watt/Volt" represents the rate of entropy generation in the magnetic field. So, recognizing the electric charge and magnetic flux as forms of energy, the SI system of units can be modified by deleting the Ampere as unneeded base unit. Such modification removes, as will be shown, redundancies in the traditional SI system of units and homogenize the units of thermal, electric, and magnetic fields. This chapter will present a study of the impacts of the new definitions of electric current and magnetic flux on proper explanation of phenomena in the field of electromagnetic and photoelectric effects, on proper understanding of the duality confusion and on the wireless power transfer that has a long history. At the end of this chapter, it will be shown how such proper understanding of fluxes leads to proper understanding of the nerve impulses and of the techniques of stimulating the neural systems for diagnosis of diseases of the neural systems.

Keywords: electric charges, magnetic flux, electromagnetic waves, entropy, thermoelectric effects, fields, quantum mechanics

1. Introduction

While we are reaping the fruits of our progress in electrical and magnetic engineering in this era, we cannot find clear definitions and units of basic entities in these sciences as the electric charges and magnetic flux. The electric current is defined in literature as a stream of charged particles, such as electrons or ions, moving through an electrical conductor or space [1]. This definition denies the fact that the measured speed of electrons in good conductors does not exceed 1 cm/second while the velocity of flow of electric current in conductors or space approaches the speed of light in addition to the impossibility of flow of electrons or any charged particles through space. Similarly, magnetic flux is defined in literature as a static region around a magnet in which the magnetic force exists [1]. This definition also ignores the meaning of the word "flux" as a part of such identity which indicates a dynamic flow of a flowable entity and not a static region of its effect.

In this study, it will be discussed analogy of laws that characterize the flow of heat, electric charges, and magnetic flux in addition to results of Faraday's experiments to show the common nature of these entities as energy in transfer by the force of their corresponding potentials [2]. The analogy of the laws governing the flow of heat and of electric charges is normally applied in simulating the thermal networks by electrical networks where the potential in both fields is measured by the same unit, Volt, and their flow is also measured by the same unit, Joule [3]. The success of Faraday in converting a ray of light into a beam of electric current when passing through an electric field and converting another ray of light into magnetic flux when passing through a magnetic field represents a proof of the common nature of all energy fluxes as EM waves driven by their corresponding potentials [4]. Faraday also discovered the phenomena of electromagnetic induction where the magnetic flux linked with a closed coil induces flow of electric charges into another coil [4]. The results of Faraday's experiments prove the similarity of natures of the electric and magnetic fluxes as convertible energies where each flux may replace the other quantitively and qualitatively. According to previously explained analogy and analysis, the electric charges and the magnetic flux can be considered as EM waves which moves independently by the force of their electric or magnetic potentials as the thermal radiation which moves independently as EM waves by the force of its own thermal potential [5].

The ammeter is traditionally defined in literature as a device whose reading is multiplied by the potential difference across a conductor to find the power passing through such conductor according to the following equation:

$$Electric\ Power\ (W) = Ammeter's\ reading\ x\ Potential\ difference\ (Volt) \qquad (1)$$

Accordingly, the unit of the ammeter's reading is Watt/Volt. Recognizing the electric charge as energy, according to the previous analysis, then the unit of its rate of flow should be Joule/sec or Watt. If the ammeter measures the electric current, traditionally defined as the rate of flow of charges, its unit should be the Watt. Such unit determines the power passing through the conductor according to the following equation:

$$\textit{Electric power } (\dot{W}) = \textit{Rate of flow of energy or elect.charges}(\dot{W}) \qquad (2)$$

Comparing Eqs. (1) and (2), we get to the conclusion that the ammeter's reading does not determine the traditionally defined electric current but another entity. Investigating the entity measured by the ammeter which has the unit Watt/Volt, it is found as a property of the electric fields (or conductors) which defines the capacity of such electric field to pass a definite rate of flow of electric energy or charges by the force of a unit potential, i.e., by 1 Volt [6]. However. this capacity is also known as function of one of the fundamental physical properties of fields which is called "Entropy" [7]. As a concluded postulate, the ammeter measures a property of the field which determines its capacity to allow the flow of a definite rate of electric charges by one unit of electric potential. According to the definition of entropy, this capacity also measures the ability of the field to destruct its exergy or to generate entropy by the measured rate [8]. So, it is possible to state that if the ammeter's reading is $\dot{S}$ Watt/ Volt, then the tested electric field has the capacity to pass $\dot{S}$ Watt by the potential difference of 1 Volt and it means also that this electric field allows the growth of its entropy or destruct its exergy by the rate of $\dot{S}$ W/Volt. The following equality may express better understanding of the entropy as fundamental property of the electric fields:

$$\begin{aligned}\text{The ammeter's reading } \dot{S} \equiv \text{ the field has the capacity to allow the flow of } \dot{S} \text{ Watt/}\\ \text{by the force of potential difference of 1 Volt}\end{aligned} \qquad (3)$$

Relying on the new definitions of the electric and magnetic fluxes as EM waves and recognizing the ammeter's reading as the rate of growth of entropy of the tested field; it was followed an exclusive entropy approach to represent all energy fluxes analytically and graphically by a modified form of the Maxwell's wave equation that replace the time in these equations by entropy [9]. The approach depends on a fact that the entropy is mainly an arrow of time [10]. Such replacement casted the Maxwell's equation into an energy frame of reference where the planes of the three coordinates of the modified eq. E (Electric Potential), H (Magnetic Potential), and S (Entropy) represent fundamental energy parameters [11]. Such representation is necessary to follow the interpretation of the devices that record the characteristics of electric charges and magnetic fluxes as E.M. waves in a frame of energy and entropy coordinates [12].

Analyzing the stopping potential phenomena of photocells in experiments that determine the Planck's constant, it was possible to measure the potential of the electric charges emitted by incident radiation [13, 14]. Such electrical potential is discovered when light of high frequency, determined by the temperature of the source of radiation according to Wien's law, falls on a metal plate or the cathode of a photocell, **Figure 1**. Electric charges were ejected from the surface of the metal plate which have definite electric potential [15]. The flow of electric charges from the cathode to the anode is found to stop when the potential of the ejected charges is resisted by an equal potential of the anode. Accordingly, the potential of the ejected charges depends on the thermal potential of the incident radiation as shown in **Figure 2** [16]. So, the

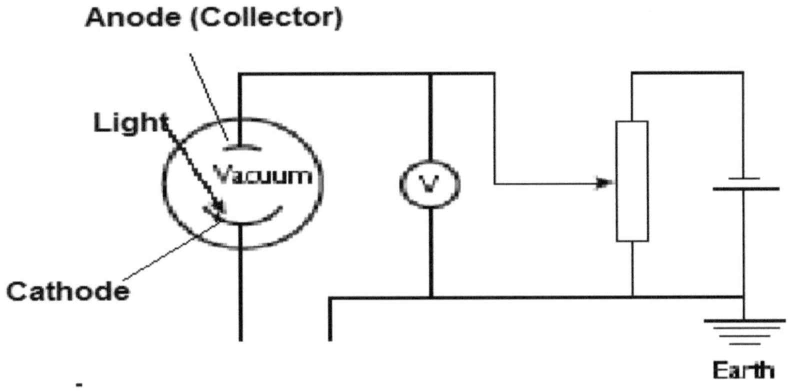

Figure 1.
The photocell experiment is designed to measure the stopping potential on the anode (collector) that stops the flow of electric current from the cathode by applying a negative potential on the anode.

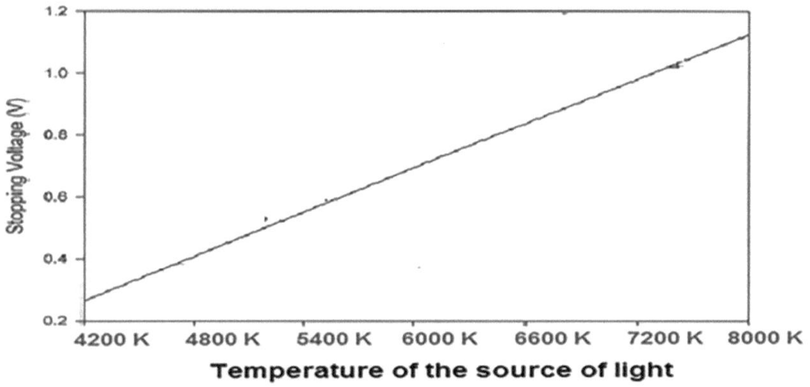

Figure 2.
Dependence of the stopping voltage in a photoelectric cell on the temperature of the source of light.

electric charges should have a potential whose polarity determines the sign of the charge, positive or negative, and its magnitude determines the strength of the source of radiation, i.e., the thermal potential or the temperature of the source of radiation.

The found slope for Sodium plate, as found in the current measurement in **Figure 1**, is found to be identical to the measured slope for potassium plate as measured by other authors [13, 15]. However, the wavelength in the abscissa of the cited references is replaced in **Figure 2** by the temperature of the source of radiation according to Wien's law of radiation in order to show the correspondence between the thermal and electric potentials [16].

The linear dependence of the electric potential of the ejected charge as a function of the thermal potential of the incident EM waves, as shown in **Figure 2**, can be considered as a base for calibrating the scale of electric potentials as function of the naturally defined thermodynamic scale of temperature. Such thermodynamic scale depends on physical laws which should not be violated [17]. However, the scale of the electrical potential has only one natural point that set its absolute zero to be defined by the earth's potential.

By analogy between electric charges and magnetic flux, it was possible to use the magnetocaloric effect (MCE) as defined in literature as a reversible temperature change of a magnetic material upon the application or removal of a magnetic field [18]. Such definition serves to find another scale for the magnetic potential in terms of the thermodynamic scale of temperature [19]. Recognizing the magnetic flux as a form of energy, the application of magnetic field on a magnetic material act as pumping or transferring magnetic flux into the material that leads to increasing its stored internal energy, and hence to increasing its magnetic potential. According to the principles of thermodynamics, the state of the magnetic material during a magnetocalorific process can be determined by its thermal and magnetic potentials, i.e., its temperature T and its magnetic strength H. So, the internal energy, as a function of the state, can be expressed as follows [9]:

$$U = U(T, H) \tag{4}$$

Accordingly, the differential of the internal energy U can be expressed as follows:

$$dU = \left(\frac{\partial U}{\partial T}\right)_H dT + \left(\frac{\partial U}{\partial H}\right)_T dH \tag{5}$$

Eq. (5) is applied in processing magnetic materials through magnetocalorific cycles which uses mutual inductions between the thermal potential (ΔT) and the magnetic potential (Δ H) for cooling of materials [19]. When applying a magnetic field (+Δ H) during a magnetizing step in the cycle, the thermal potential will increase, allowing the magnetic material to be cooled by surroundings at normal temperature. Then the removal of the applied field as demagnetizing step of the cycle will tend to a subcooling effect ($-\Delta$T) which can be used for cooling of any medium. The cooling step ends the magnetocalorific cycle where the magnetic material will return to its initial state. So, the net change of internal energy of the magnetic material, as a function of state, will be zero, i.e.,

$$dU = 0 \tag{6}$$

Substituting in Eq. (5), we get:

$$\left(\frac{\partial U}{\partial T}\right)_H dT = -\left(\frac{\partial U}{\partial H}\right)_T dH \tag{7}$$

Defining the thermal and magnetic capacities $C_{thermal}$ and C_{mag} of the magnetic material by the following equations:

$$C_{Thermal} = \left(\frac{\partial U}{\partial T}\right)_H \tag{8}$$

and,

$$C_{mag} = \left(\frac{\partial U}{\partial H}\right)_T \tag{9}$$

Substituting in Eqs. (8) and (9); the following relation between dH and dT will be:

$$dH = -\frac{C_{Thermal}}{C_{mag}} dT \tag{10}$$

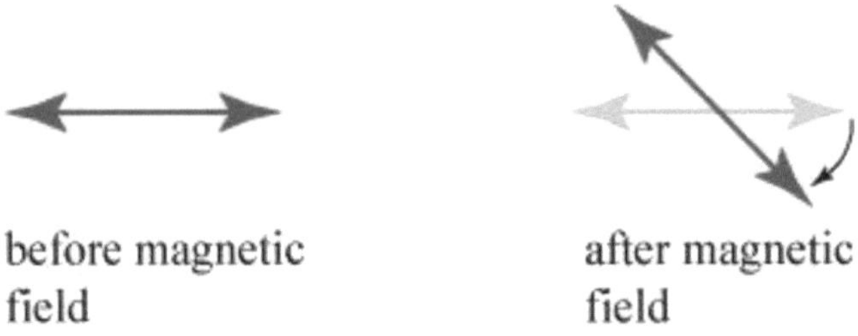

Figure 3.
Rotation of a light ray by a magnetic field [4].

The experimental verification of Eq. (10) is seen in **Figure 3** which shows the influence applying magnetic field of specific intensity "H" on the rise of temperature of magnetic materials, "T" known as the magnetocaloric effect [20]. Such measurement results can be considered also as a method of calibrating the magnetic potential, according to the thermodynamic temperature scale as reference potential, which is defined by the magnetocaloric effect, Eq. (10) [21]. Accordingly, all kinds of energy in transfer, as thermal or electric or magnetic energies may have the absolute temperature scale as a reference or a unique scale of measurement for their potentials, or the electric and magnetic potentials.

This chapter will also show the impacts of the introduced understanding, as proper understanding, of the natures of the electric charges and magnetic flux on creating a modified SI system of units that removes redundancies in the traditional SI system. However, proper understanding of electric charges and magnetic flux will lead, as will be shown, to a proper understanding of the thermoelectric and the photelectric effects. Similarly, the new understanding will help in adopting the wireless power transmission starting from Nikola Tesla tower to the long-distance wireless power transfer by beamed microwaves and the magnetic resonance. Such proper understanding may also save the neurologists from their blunder in understanding the proper interpretation of nerve impulses as electric charges moving independently by the force of their associated potentials. The reprint of their stimulating devices assures the neural networks perform normally as electric networks whose nerve impulses move as electrified wave packets of energy determined by the integral: $\int_0^\lambda \boldsymbol{E}\, \boldsymbol{dS}$, where λ is the wavelength of the nerve impulse as an EM wave and E is its electric potential.

2. Faraday's experiments as an approach for a common understanding of the energy fluxes

Faraday held many experiments whose results made him convinced that he had succeeded in magnetizing and electrifying a ray of light and in illuminating a magnetic line of force [4]. He postulated that such identities are so directly related and mutually dependent that they are convertible, possess equivalents of power in their actions, and have a common origin. Faraday wrote this postulate as the introduction to the nineteenth series of his "Experimental Researches in Electricity" [4]. The influence of the magnetic field on the beam of light that Faraday observed is now known as Faraday's rotation because his experiment illustrated the rotation of a ray of light by the action of magnetic field, whatever the magnet's configurations are, as seen in **Figure 3**. This was the first definitive evidence that light and electromagnetism are related [4].

DOI: http://dx.doi.org/10.5772/intechopen.106962

According to his successful experiment of electromagnetic induction, **Figure 4** [22], Faraday succeeded in 1831 to lay the groundwork to James Clerk Maxwell to formulate his own electromagnetic theory of light, predicting that both light and radio waves are electric and magnetic phenomena and led to important inventions such as electric motors, transformers, inductors, and generators [23, 24].

In his theory of EM waves, Maxwell predicted that EM disturbances traveling through empty space have electric and magnetic fields at right angles to each other and that both fields are perpendicular to the direction of the wave as a time coordinate [25]. He also concluded that the waves move at a uniform speed equal to the speed of light and that light is a form of the EM waves. So, EM waves are described as simultaneous oscillating motions of electric and magnetic perpendicular waves, **Figure 5** [26].

Maxwell described the electric and magnetic fields arising from the propagation of the EM waves by the following equations as a standard wave form [1]:

$$\left(\nabla^2 - \frac{1}{c^2}\frac{\partial^2}{t^2}\right)\mathrm{E} = 0 \tag{11}$$

$$\left(\nabla^2 - \frac{1}{c^2}\frac{\partial^2}{t^2}\right)\mathrm{H} = 0 \tag{12}$$

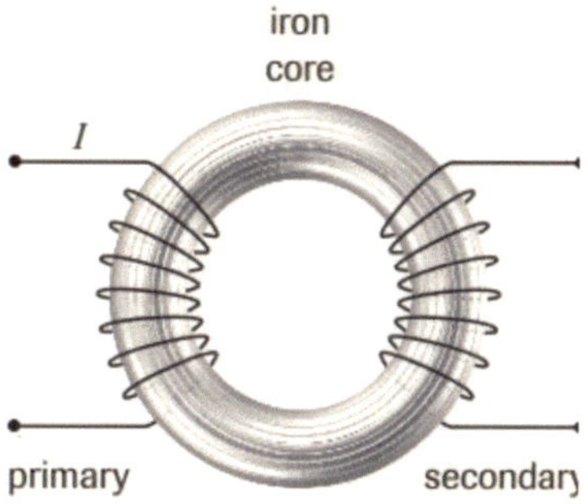

Figure 4.
Faraday's induction coil where the flow of electric current in the primary coil induces magnetic flux in the iron core, and the core's magnetic flux induces electric current in the secondary coil [22].

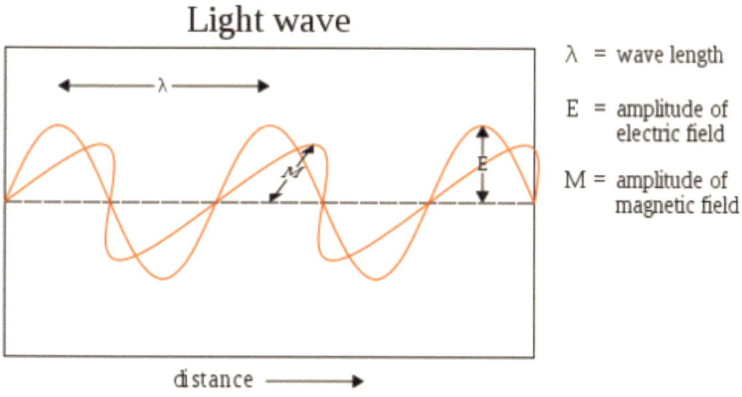

Figure 5.
Maxwell's description of EM waves as simultaneous oscillating motions of electric and magnetic perpendicular waves [1].

3. Entropy approach for proper understanding of the fluxes

Literature that dealt with thermodynamic systems are concerned mainly by transfer of mechanical and thermal energies [27]. Similarly, literature which dealt with flow of electric or magnetic energies were concerned mainly by solids which have mainly electric or magnetic interactions [28]. Both analyses are ineffective when dealing with a general thermodynamic system that may involve fluid flow while it is subjected to transfer of heat, electric, magnetic, and mechanical energies. However, literature that dealt with the transfer of electric and magnetic energies gave a narrow space to association of the energy transfer by flow of entropy as an effecting energy property in any physical system [29]. Such truncations paralyze explanation of different electromagnetic phenomena where the entropy is the key to the general analysis of any energy interaction, transformation, and conservation [2].

The flows of energy in physical systems, in general, are driven by forces in conjugate pairs [30]. So, the heat flow is driven by differences in temperature between the system and the surrounding, volume flow is driven by differences in pressure, electric charge flow by differences in electrical potential, magnetic flux by difference in magnetic potential, and mass flow by differences in concentration. According to literature, the flow of heat is associated mainly by production of entropy and the total rate of entropy production, $\sum \boldsymbol{S}_i$, is not limited by equilibrium theory. According to analogy between heat, as thermal energy, and electric and magnetic energies, electric and magnetic energies should be also associated by entropy production according to the equation [30]:

$$\dot{S}_{tot} = \sum p_i J_i \tag{13}$$

Where $\boldsymbol{J}_i$ represents an energy flux associated by a driving force, or potential $\boldsymbol{p}_i$. However, such entropy production is equal to the destruction of the exergy of the systems or fields [30]. Accordingly. it was introduced a fundamental equation for a system embracing thermal, electrical, and magnetic energy interactions as follows [2]:

$$dU = T\,dS + E\,d\mathrm{S} + H\,dS - P\,dV \tag{14}$$

Eq. (14) expresses the flows of electric and magnetic energies as a potential force times a differential of the corresponding increase entropy of the system whose integrals can be expressed in property diagrams between the entropy and the electric or magnetic potentials of the system [2]. The differentials of entropy are exact differential as the entropy is a property of the system. Hence, Eq. (14) can be considered as a fundamental equation as it involves only properties of the system [2].

Maxwell had succeeded in describing the flow of thermal radiation by the standard wave form represented by Eqs. (11) and (12). For a visualizing the natures of the electric charges and magnetic flux, it is introduced an entropy approach that casts the Maxwell's wave equations into an energy frame of reference through replacing time by entropy as a unique function of time as follows:

$$\left(\nabla^2 - \frac{1}{\mathrm{c}^2}\frac{\partial^2}{\mathrm{s}^2}\right)\mathrm{E} = 0 \tag{15}$$

$$\left(\nabla^2 - \frac{1}{\mathrm{c}^2}\frac{\partial^2}{\mathrm{s}^2}\right)\mathrm{H} = 0 \tag{16}$$

Such modification of the Maxwell's equations enables the representation of all energy fluxes in planes formed by the three energy related coordinates found in the modified Maxwell's equation: E (Electric Potential), H (Magnetic Potential), and S (Entropy).

Figure 6 shows the graphical representation of the modified wave equations. In this figure, the imparted electric energy per wave of length λ_s is represented by the swept blue area of the EM wave in the E-S plane and the magnetic energy per wave of length λ_s is represented by the red swept area of the EM wave in the H-S plane. The sum of the two areas represents the imparted energy by one EM wave of wavelength λ_s. This energy is denoted by the symbol "Ћ" and has the unit Joule/cycle and can be calculated by the following equation [31].

$$Ћ = \int_0^{\lambda_s} |E\, dS_{elect}| + \int_0^{\lambda_s} |H\, dS_{mag}| \tag{17}$$

λ_s is the wavelength in the entropy coordinate, as an arrow of time, which is determined by the measured rate of entropy flow according to the following equation: [2]?

$$\dot{S} = \lambda_s\, x\, \nu \tag{18}$$

ν represents the frequency of the EM wave. Denoting the amplitudes of the energy waves in the electric and magnetic planes as $\mathbf{E^0}$ ***and*** $\mathbf{H^0}$, and considering a form of sinusoidal wave for both energies, the imparted energy per wave according to Eq. (18) can be found as follows:

$$Ћ = \int_0^{\lambda_s} |E^0 \sin s\, dS_{elect}| + \int_0^{\lambda_s} |H^0 \sin s dS_{mag}| \tag{19}$$

Mathematically, such integrals is estimated as follows [31]:

$$Ћ = 1.414\, \lambda_s (E^0 + H^0) \frac{Joule}{cycle} = 1.414\, \frac{\dot{S}}{\nu} ((E^0 + H^0)) \frac{Joule}{cycle} \tag{20}$$

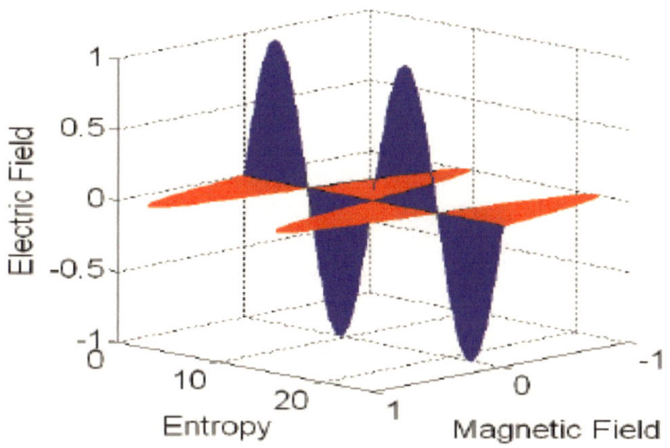

Figure 6.
Flow of electromagnetic waves in an energy frame of reference that show the electric energy flow in an electric field-entropy plane and the magnetic energy flow into a magnetic field – Entropy plane [2].

Hence, the power of this wave is calculated as follows:

$$\dot{W} = \hbar.\nu = 1.414.\dot{S}.\left(E^0 + H^0\right) \text{ Watt} \tag{21}$$

According to Eq. (21), the EM waves flow in wave packets at a rate of energy flow: $\hbar$ x ν Watt. Eq. (20) shows the energy packet per wave $\hbar$ is proportional to the sum of the wave's amplitudes of the electric and magnetic components of the EM wave and to the capacity of the medium that allows the flow of the power $\left(\boldsymbol{E^0} + \boldsymbol{H^0}\right).\dot{\boldsymbol{S}}$ Watt by the force of 1 Volt.

According to Eq. (21), the energy of the EM wave packets does not depend on the frequency of the EM waves but on properties of the medium and amplitudes of the wave. Such proved conclusion violates Planck's postulate that the energy of the wave packets, or photons, is proportional to the frequency of the EM wave and the claim that these packets behave as particles which can eject electrons to produce electricity [31]. Such statement was postulated by Einstein and was violated by S. AbdelHady as the momentum of the claimed energy particle is negligible if compared to the momentum of an electron [32]. However, the term "h ν" should not identify a quantity of energy as it contains the frequency "ν" that has the dimension $1/_T$ which defines a rate of flow. However, according to the results of Planck's experiment in **Figure 2**, the rate of flow of energy is found to be the most effective parameter in the thermoelectric and photoelectric effects and it is not the quantity of energy [33]. The Einstein's postulate of dual natures of the wrongly defined electric current as electrons and his explanation of the ability of the EM waves to eject electrons paved a way to the quantum mechanics and the uncertainty principles of Heisenberg [34]. However, S. AbdelHady found an entropy approach that depends on Boltzmann's principle is more realistic than the quantum approach that depends on Schrodinger's equations of imaginary terms and leads to virtual, or unreal, solutions [35, 36]. Additionally, Shannon found, through another entropy approach, a more adequate uncertainty measures than the Heisenberg's uncertainty measures of the indefinite quantum distributions [37].

The followed entropy approach in the presented study realizes the role of entropy as a main property of thermal, electric, and magnetic fields that characterize the flow of such energies though the corresponding fields. Casting Maxwell's wave equations into the E-H-S frame made it possible to represent the positive or negative charges as energy of positive or negative electrical potentials. **Figure 7** represents a positive electric charge of positive potential where the "E" coordinate of the entropy axis is not at zero potential, but it has a positive value "+Δ E" which determine the charge's polarity. The mathematical formulation of such wave is written as a solution of the modified Maxwell's equation that considers the wave starts by initial positive potential Δ E at τ = 0, that corresponds to the zero time and the zero entropy, i.e., at s = 0, as the entropy is a direct function of time [39].

Assuming the following initial condition:

$$E_{\tau=0\equiv s} = \Delta \text{ E} \tag{22}$$

So, the solution of the modified Maxwell's wave equations will be as follows [38]:

$$\text{E (r, s)} = g_1 \text{ (}\omega \text{ s–k r)} + \Delta \text{ E} \tag{23}$$

$$\text{H(r, s)} = g_2(\omega \text{ s–k r)} \tag{24}$$

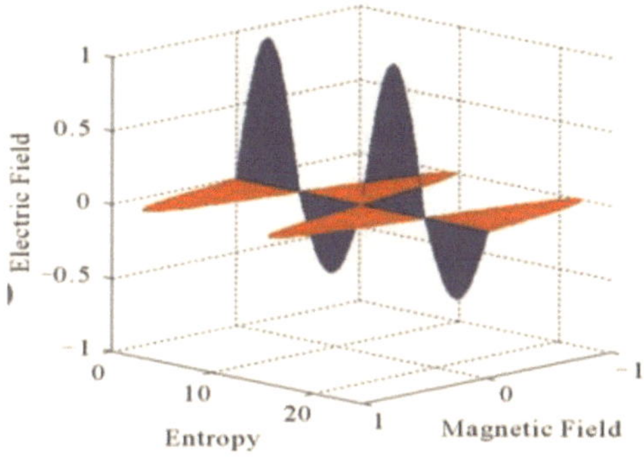

Figure 7.
Graphical representation of a positive electric charge distinguished by the oscillation around an entropy axis whose E coordinate "electric field" has a positive value + Δ E [38].

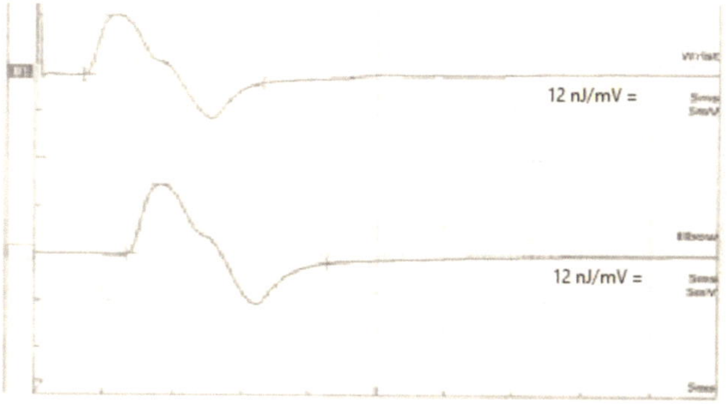

Figure 8.
A plot of the stimulating signals from a stimulation device as an E-s record of the signal. [hospital plot].

Similarly, the negative electric charge can be represented by EM wave whose electric wave has a negative potential - Δ E. Additionally, the magnetic flux can be also represented by EM wave whose magnetic wave has positive or negative potentials +/− Δ H and represented analytically by the following equation [40].

Figure 8 shows a stimulator record of a stimulus charge used in neural diagnosis recorded as a wave oscillating about an axis of positive potential ordinate, i.e., it is a positive charge. The abscissa records the growth of the entropy of the field, or in tissues of the body, during the flow of the charge in n. joule/volt, and the ordinate records the variation of the potential of the injected charge during the process in Volts. According to the data on the record, it is possible to read the recorded growth of entropy during one wave = Δ S = 3 nano-Joule/volt and the recorded time duration of the pulse = 0.2 milli Second.

However, the rate of growth of entropy inside the body tissues during stimulation process represents a property of such tissues that may help in the diagnosis or therapy processes.

$$\dot{S} = \frac{\Delta S}{\Delta \tau} = \frac{3\,x\,10^{-9}}{0.2\,x\,10^{-3}} = 15.\ 10^{-6}\ \text{Watt/Volt}. \tag{25}$$

Such capacity of the neural tissues to pass high power by the force of 1 Volt proves that the neural axons are very conductive and prove the ability of the neural cells to pass electric current. Additionally, it is a proof that the electric current is not flow of electrons, otherwise, it will not pass through the neural cells. This result also proves the nature of electric current as EM waves that have electric potential as represented by the stimulator record in **Figure 8**.

Reading the data of the involved electrical potential of the resource of electric power during the stimulation process was found = 5 Volt. Hence, the rate of flow of electric charges into the tissues during the charge pulse is calculated as follows:

$$E.\dot{S} = 5 \times 15.10^{-6} = 75\ \mu\text{Watt}. \tag{26}$$

The similarity of the stimulator records in **Figure 8**, as an actual record of the stimulating charge in neural system, and the theoretical solution of the modified Maxwell's equation in **Figure 7** represents the evidence of the truth of the introduced entropy approach for representing the electric charges in the E-s plane. Other faces of truth of the new definition of electric charges as EM waves is found in: a). similarity of the shape of the EM wave as a sinusoidal wave and the plotted shape of the wave also as a sinusoidal wave; b). the unit of measurement of the rate of growth of entropy as found by the ammeter's readings in joule/volt and as found on the abscissa of the measurement record in n. Joule/sec; and c). the measured electric potential of the charge as identified by Maxwell' s equation in Volts and, as plotted on the ordinate of the machine record in Volts too. So, the ammeter does not measure a rate of flow of electrons that should be expressed by a mass flow rate or by units of power, but it is an energy related parameter that is represented in an energy frame tied by the rate of entropy growth.

4. Impact of the proper understanding of fluxes on proper modifications of the SI system of units

As a matter of fact, declared by many authors, it is impossible to accept the found redundancies in the SI system of units [41]. According to the proper understanding of the electric charge as energy measured by the Joule, its rate of flow should have the unit "Watt." Additionally, the ammeter's reading as previously discussed has the unit Watt/Volt. Hence, there is no need to the "Ampere" which does not define the ammeter's reading, it is a unit of unidentified entity, and is wrongly defined as the rate of flow of electric charges. So, modifying the SI system by deleting the Ampere as a meaningless basic unit represents, as will be shown, a successful modification step. So, the seven basic unit of traditional SI system of units will be reduced to six units only. As another important result of recognizing the natures of electric charge and magnetic flux as forms of energy, it is the previously found dependence of the scales of their potentials on the thermodynamic temperature scale which is determined by fundamental laws of thermodynamics [42]. However, the "Volt" is also selected as a basic unit for all potentials where the Volt is already used in measuring the electric and thermal (by using thermocouples) potentials. The induction transformer can be also used to define the mutual dependence between the electric and magnetic potentials and the result of the experiments of determination the Planck's constant, **Figure 2**, can be used to define the dependence of the electrical potential scale on the thermodynamic temperature scale [43, 44]. However, the earth represents the "Zero" reference point of the scale of electric potential due to its high capacity. Finally, the base units of the modified SI system will be stated as follows:

Basic Units	Dimension	Unit
Length	L	m.
Time	T	sec.
Mass	M	kg
Amount of Substance	N	Mole
Energy Potential	**P**	Volt
Luminous intensity	J	cd

Accordingly, the derived units in the modified SI system would have the following dimensions and units:

Dérivéd Unit	Dimension	Unit
Velocity	$L\,T^{-1}$	m/sec
Accélération	$L\,T^{-2}$	m/sec^{-2}
Force	$M\,L\,T^{-2}$	Newton
Pressure	$M\,L^{-1}\,T^{-2}$	Pascal
Energy (All Fluxes)	$M\,L^{2}\,T^{-2}$	Joule
Power (All Fluxes)	$M\,L^{2}\,T^{-3}$	Watt
Resistance (All Fields)	$M^{-1}L^{-2}T^{3}V$	Ω
Conductivity (All fields)	$M\,L\,T^{-3}\,V^{-1}$	Watt/V. m

The units in the scope of the thermal, electric, and magnetic energies are as follows: The thermal potential T, and the electrical potential E, and the magnetic potential or the magnetic field intensity H have the same dimension "P" and all are measured by the same unit "Volt."

The heat $Q_{thermal}$, The electric charge $Q_{electricpal}$, and Magnetic flux B have the dimensions "$M\,L^{2}\,T^{-2}$" and measured by the same unit "Joule."

All powers as the rate of flow of energy have the dimensions "$M\,L^{2}\,T^{-3}$ and are measured by the same unit "Watt."

All resistances (or Reluctances) have the same dimensions as "$V\,M^{-1}\,L^{-2}\,T^{3}$ and the same unit "Ω."

Many conflicts are found in the traditional literature when searching the SI units and dimensions of resistance, potentials, and permeability of different energy fields [1]. However, the presented modified system realizes the homogeneity of the units of the analogical parameters of different fields, i.e., the analogous parameters in the energy field have the same units and dimensions. Such result confirms the convertibility and unity of the origin of the three known forms of energy as Faraday described and expected [4].

As a comment on the versality of the modifications, S. Abdelhady discovered by using such homogenous system of units a dimensional error in Ampere's law [45]. Comparing the dimensions of the R.H.S. and L.H.S. of Ampere's law that estimates the electromotive force "E" resulting from the time-rate of change of magnetic flux "((dB)/(d t))"crossing a specific area "A" which reads:

$$A.\frac{d\,B}{d\,t} = -E \tag{27}$$

Such equation is dimensionally incorrect as the dimensions of both sides of the equation are unequal. To correct the dimensions of this equation such that the dimensions of both sides of this equation would be similar, the L.H.S. of the equation should be multiplied by the Reluctance of the magnetic field or the magnetic resistance of air as follows:

$$A\ \mathrm{m}^2.R_{mag}\frac{Volt}{Joule}\frac{d\,B\ \frac{\mathrm{joule}}{\mathrm{m}^2}}{d\,t\ \mathrm{sec}} = -E\ \mathrm{Volt} \tag{28}$$

In this way, the Ampere's equation is dimensionally correct where its error is corrected by adopting the modifications introduced by the proper definitions of electric current and magnetic flux as forms of energy of analogical potentials.

Additionally, it is possible to use the new definition of the electric charge as energy to find a relation between the "Volt" as a selectable unit of measuring the electric potential and the "Newton" as a determined unit of measuring the force according to Newton's law of motion. According to literature, the following equation determines the force applied on a suspended electric in an electric field as follows [1]:

$$F = Q.E \tag{29}$$

Recognizing the nature of the electric charge as energy of electric potential, it is possible to determine the force between two charges A and B according to considering the electrical force like the gravity force that is inversely proportional to the distance between the charges while keeping the dimensional homogeneity as follows:

$$F = \frac{1}{k}\sqrt{\frac{[(Q_aE_a)(Q_bE_b)]}{r^2}}\ \mathrm{Newton} \tag{30}$$

If the two charges have the same energy and potential, the force between them will be:

$$F = \frac{1}{k_e}\frac{Q\,E}{r} \tag{31}$$

Substituting the units of the involved parameters in Eq. (31) by units of the modified SI system of units, the dimensions of the constant of k_e is found to be the "Volt." So, its value may be selected to fit the correct relation between the fixed scale of force and a selected scale of the electric potential. However, the force determined by Eq. (31) may be repulsive or attractive force according to the sign of both charges.

According to the similarity of the electric and magnetic fields, it is possible to also expect a similar equation that determines the force of attraction, or repulsion, between two identical magnetic identities each of magnetic flux "B" and magnetic potential of "H" to be:

$$F = \frac{1}{k_m}\frac{B.H}{r}\ \mathrm{Newton} \tag{32}$$

Substituting the units of the involved parameters in Eq. (30) by units of the modified SI system of units, the dimensions of the constant of k_m is found to be the "Volt." So, its value may be selected to fit the correct relation between the fixed scale of force and a selected scale of the magnetic potential.

5. Impact of the proper definitions of the fluxes on proper understanding of thermoelectric and photoelectric effects

Recognizing the electric charge as energy of electrical potential, it is possible to find plausible explanations of newly discovered phenomena in the fields of thermoelectric and photoelectric effects [46–48]. It will be summarized in this section such plausible explanations.

Considering the thermocouple in **Figure 9**, it is constructed of two different metals "A" and "B." If its junctions are placed into two heat reservoirs "1" and "2" where the difference in temperature between them is "Δ T," defined as $\boldsymbol{\Delta T} = \boldsymbol{T}_2 - \boldsymbol{T}_1$, 1, then an open circuit voltage, "Δ E" will be obtained between the two junctions [28]. The measured output electric potential is found proportional to the temperature difference between the junctions of the two conductors A and B according to the following Seebeck equation [1]:

$$\Delta E = \propto_{AB} \Delta T \tag{33}$$

where "$\propto_{AB}$" is the relative Seebeck coefficient, between the conductors A and B, expressed in μ V/K. This coefficient depends mainly on the choice of the two materials used in the thermocouple and the temperature of the junction at the higher temperature "T_{max}". The magnitude of the relative Seebeck coefficient of the junction between the two materials or metals A and B can be evaluated as the difference between the Seebeck coefficient of the two metals as follows [1].

$$\propto_{AB} = \propto_B - \propto_A \tag{34}$$

The direct relation between the produced electric potential and the difference of the thermal potentials between the two junctions plays the main role in the use of thermocouples in temperature measurements and in thermoelectric generators.

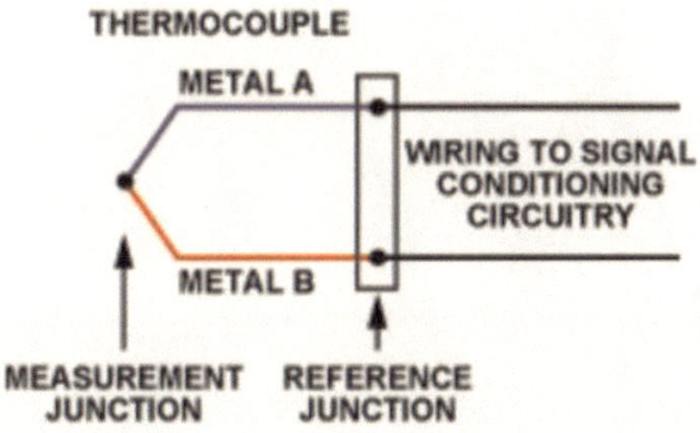

Figure 9.
A thermocouple of two different materials "A" and "B." that has a reference junction and a measuring junction [1].

However, there is a relation between the Seebeck coefficient "α" and the energy bandgaps of materials "E_g" which was found by Goldsmid and Sharp as follows [49]:

$$E_g = 2\,e\,|\alpha_{max}|T_{max} \tag{35}$$

Where e is the electron's charge = $1.602.10^{-19}$ Joule. Eq. (35) signifies a relation between the Seebeck coefficient and the energy- bandgaps of materials of the junction that characterize the transitional effect from thermal potential "ΔT" to electric potential "ΔV." The tables of Seebeck coefficients of materials and the tables of its energy bandgaps indicate a direct relation between these two physical properties [15].

Recognizing the flow of electric charges as a flow of EM waves of electric potential, it is possible to interpret the transition of the thermal potential of the flowing heat "ΔT" across the thermocouple junction into electric electrical potential "ΔE" to be done by the Seebeck effect which is expressed by Eq. (33). Such transition can be explained as exchanging the thermal potential of the input heat "ΔT" by electric potential "ΔV". Such transition occurs when the input radiation crosses the junction between the materials "A" and "B" of different bandgaps into electrical potential by the force of a thermopower resulting from the difference of temperatures and the difference of bandgaps of the two materials of the junction [50].

However, the emf produced in the operations of thermocouples, as a tool to measure the temperature, is usually very low. To increase the voltage difference, it is used thermopiles where several junction-pairs are connected in series as shown in **Figure 10**.

As the thermocouple junction pairs are placed in series between a source of heat at high temperature "T_h" and a heat sink at low temperature T_l,as shown in **Figure 11**, then the output voltage can be estimated as tsum of the gained electric potentials during the flow of the electromagnetic waves of thermal potentials in a unique direction across the successive junctions. So, their thermal potentials, ($T_h - T_l$), will be converted by the Seebeck effect into electric potentials which will be accumulated in as the sum of the unit potential [51]. So, it is possible to estimate the total gain of electric potentials as the sum of these individual gains at successive junctions as follows:

$$\Delta V = \propto_{AB}(T_h - T_l) + \propto_{BA}(T_l - T_h) + \propto_{AB}(T_h - T_l) + \propto_{BA}(T_l - T_h) + \propto_{AB}(T_h - T_l) + \dots . \tag{36}$$

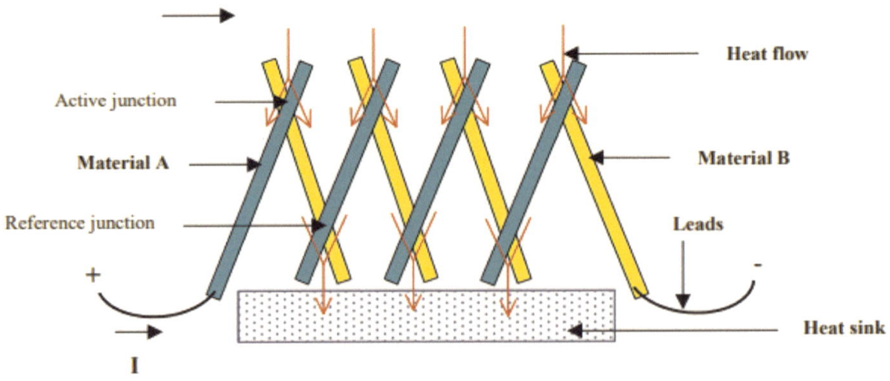

Figure 10.
A thermopile made of thermocouple junction-pairs of two metals connected electrically in series to increase generated EMF [51].

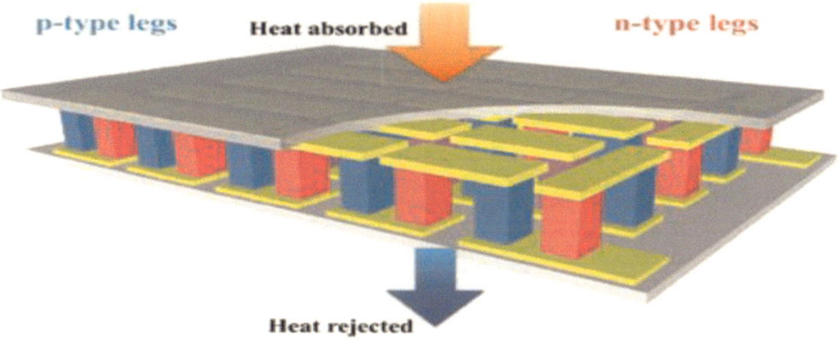

Figure 11.
A schematic diagram of a thermoelectric generator where the p-type and n-type legs have been bonded together as seen by upper and lower conducting plates to accumulate the gained electrical potentials of the output electric current [15, 52].

Substituting the Seebeck according to Eq. (34) as:

$$\propto_{AB} = \propto_B - \propto_A, \propto_{BA} = \propto_A - \propto_B = -\propto_{BA}, (T_h - T_l) = -(T_l - T_h) \quad (37)$$

So, the total accumulated electrical potential will be:

$$\Delta V = \sum [\propto_{AB}(T_h - T_l)] = n \propto_{AB}(T_h - T_l) \quad (38)$$

Eq. (38) indicates that the electric potential of a thermopile is duplicated by the number of the used junctions. Such equation is applied in measurement of temperature and in thermoelectric generators for predicting the amplified electric potential. Without recognizing the electric charge a EM wave of electric potential, we cannot find a plausible explanation of accumulation of the electric potentials of such energy when crossing successive junctions and measuring its potential according to Eq. (38).

A classical TEG is shown in **Figure 11**. According to recognizing the electric current as a flow of EM waves that have electric potential, it is possible to conclude the operation of the shown TEG module would follow the same principles of operation of the thermopile in **Figure 10**. As the junction pairs in this module are connected in series; then the output potential of the electric current will be similarly estimated as the sum of the gained electric potentials across the successive junctions by the Seebeck effect. So, if the flow of electric current is defined as flow of electromagnetic waves of accumulating potentials, it is possible to estimate the total gain in the electric potential as the sum of these individual gains at successive junctions as found by Eq. (38).

The Solar cells or photovoltaic "PV" cells are known as tools to convert solar energy into electricity by the PV effect. Such PV effect was traditionally defined in literature of physics as bouncing electrons by the incident photons of light across the cell's junction [1]. Such definition was found to violate the principle of conservation of momentum where the momentum of any incident photon is negligible when compared to the momentum of any orbiting electron in an atom [53]. Additionally, such explanation does not offer plausible explanations in case of multi-junction thin films or organic solar cells as such electrons cannot jump freely from one junction to hit electrons in another junction and the fact that electrons cannot flow through organic tissues in case of organic PV cells [54]. However, the gained potential in the PV cells

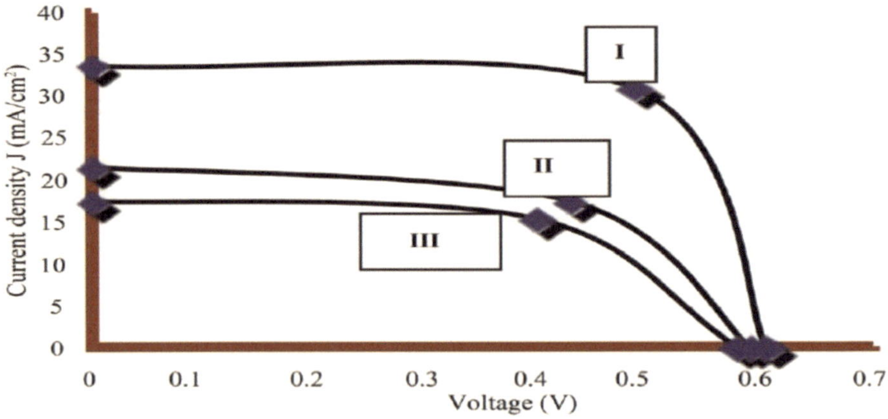

Figure 12.
Typical I-V characteristics of an illuminated single crystal silicon solar cell at different values of solar radiation 100 mW/cm² (I), 60 mW/cm² (II) and 40 mW/cm² (III) respectively.

depends on the difference of the energy bandgaps of the junction's materials, i.e., the difference between bandgaps of semiconductors of free electrons and semiconductors of holes [3]. As Eq. (33) states a direct relation between the energy bandgap and the Seebeck coefficient of materials, it was possible to also postulate the conversion of the incident radiation of thermal potential into EM waves by Solar cells into electric current or EM waves of electrical potential by the Seebeck effect after crossing the p-n junction between two dissimilar materials of different Seebeck coefficients [55]. Truth of this postulate can be proved by investigating the measurement of the open voltage potential of output current from a single crystal silicon solar PV cell under variable intensity of incident solar radiation, shown in **Figure 12**.

According to the shown dependence in **Figure 12**, the open voltage potential is affected only by the temperature of the source of radiation, the sun, while the intensity of solar adiation influences the current density [43].

According to the seen results, the open voltage potential, V_{oc} = 0.7 Volt is found to be affected only by the temperature of the source of radiation, T_{sun} = 6100 K. Such potential is found as a Seebeck effect according to Eq. (31) by substituting the value of the relative Seebeck coefficient between the materials of the p-n junctions, $\propto_{pn}$ according to table of the Seebeck coefficients and substituting the incident thermal radiation has the thermal potential = 6100 K.

$$\text{So, } (T_{sun} - T_{env}) = 6100\text{–}300 = 5800\,K. \tag{39}$$

$$\text{Hence; } \Delta V = \propto_{pn} = 6100\text{–}300 = 5800\,K.$$

$$\text{Hence; } \Delta = \propto_{pn}.\mathbf{6100} \approx \mathbf{0}.7\ \text{Volt} \tag{40}$$

Such results indicate the truth of defining the photovoltaic effect as a Seebeck effect and the nature of electric charges as flow of EM waves of definite electrical potential. However, Planck postulated that the energy of light moves in packets whose magnitude is proportional to the frequency where the constant of proportionality the Planck's constant "h" [56]. As previously explained, the effective parameter in the thermoelectric and EM wave's interactions is the power of the flow which depends on the frequency not on the quantity of energy [57]. As previously discussed, recognizing the definition of the electric charges as EM waves ends the duality confusion and the quantum imagination [58].

Considering I-V performance of triple junction solar cell that consists of successive layers of GaInP, GaAs and GaInNAs and the performance of four junctions – solar cell consisting of successive layers of GaInP, GanAs and GaInNAs and Ge stacked on top of each other [59]. It was possible to prove the validity of such PV phenomena as a Seebeck effect according to applicability of Eq. (38) for calculating the accumulated potentials of the output electric power [60]. The measured potential of the output current from multi-junction solar cell is found to be the sum of the converted thermal potentials into electric potential at the successive junctions [61]. The estimated efficiency of such multijunction cells according to Eq. (38) fits the measurement results that exceeds the found limit by Shockley and Queasier which was wrongly determined by considering the flow of electric charges is a flow of electrons [62].

6. Impacts of proper understanding of energy fluxes on plausible explanation of wireless power transmission

Recognizing the natures of electric and magnetic flux as EM waves that have electric or magnetic potentials, the electromagnetic power transmission looks like a viable natural process for the transfer of power through space. In a wireless power transmission system, a transmitter device, driven by electric power from a power source, generates a time-varying electromagnetic field, which transmits power as EM waves of electric or magnetic potential across space to a receiver device. The receiver device receives EM waves that have a high electric potential from the field and supplies it to an electrical load [63]. The technology of wireless power transmission is also a key to increasing the mobility, convenience, and safety of electric and electronic devices for all users.

Wireless power techniques that depend on such EM nature of all energies fall into two categories, near field and far-field. In near field or non-radiative techniques power is transferred as EM waves, as have been defined, over short distances through magnetic fields that uses mainly inductive coupling between coils of wire [8]. In far field or radiative techniques, also called power beaming, power is transferred either as energized EM waves in the form of microwaves done by magnetron, or as Excited EM in the form of Laser beams as done by laser systems [64]. These techniques can transport energy longer distances but must be aimed at the receiver. Proposed applications for this type are solar power satellites, and wireless powered drone aircraft [65].

However, the magnetron is a cross-field device which uses an electric field in conjunction with a magnetic field whose field-energy lines are at right angles to each other [66]. Accordingly, the magnetron pumps electrical and magnetic potentials into the EM waves passing through it which increases their amplitudes E^O and H^O to produce high energy wave packets. As the power of the flowing microwaves or micro-EM waves equals the product of such high energy per wave times the high frequency of the microwaves, the result will be extremely high-power wave packets, or photons, according to Eq. (38) which reads:

$$\dot{W} = \hbar.\nu = 1.423.\dot{S}.(E^0 + H^0) \text{ Watt} \tag{41}$$

Such beam of high-power packets is used also in the microwave ovens to heat food where the energized EM waves are sent through the cavity of the oven to fill its

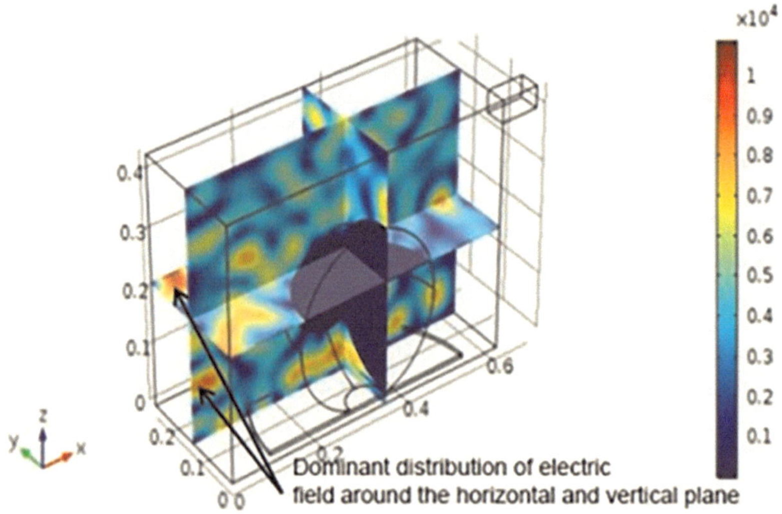

Figure 13.
Electric field distribution (V/m) of a microwave oven. The applied input power P = 500 W and time 60 sec. [66].

volume with extremely energized wave backets flowing at extremely high frequency. **Figure 13** shows the glowing packets of energy inside the microwave oven interpreted as electric or magnetic charges. Such high energized backet with extremely high potentials at high rate of flow cooks the food in extremely short time if compared to normal electric heating by Joule' s effect.

7. Impacts of proper understanding of energy fluxes on true understanding of nerve impulses

Neurologists find during their practice the nature of nerve impulses as electric signals identified by flow of charges at high speed and a measured electric potential and they measure the ability of the neural tissue to allow flow of electric current [67]. However, the traditional definition of electric current as a flow of electrons forced them to abandon such directly identified nature of the nerve impulse as electric signals due to believing the impossibility of electron to flow through organic nerves [68]. As a result of such confusion, the nerve impulse is traditionally defined in their literature as an ionic current that is moving along the length of the neurotransmitters by the force of an unknown "action potential" and they imagine a separate motion of such action potential. Such confusion can be ended by a proper understanding of the nature of the observed electric signals as a flow of electric charges properly defined as EM waves which have a definite electric potential [69]. Such innovation was discussed in a previous article by S. AbdelHady where the action potential is defined as the electric potential of the electric signal emerged from the brain as seen in **Figure 14** [69]. Such electric signals of assigned electric potential are created in the brain which accounts for 25% of the body's total glucose utilization and 20% of oxygen consumption [70]. It is a metabolically "demanding" organ with intense heat production [71]. It was discussed in a previous study the conversion of the heat produced electric to signal of action potential in the neural cells of the brain by the thermopower of such heat [50]. Such power is activated by the temperature difference across the cell

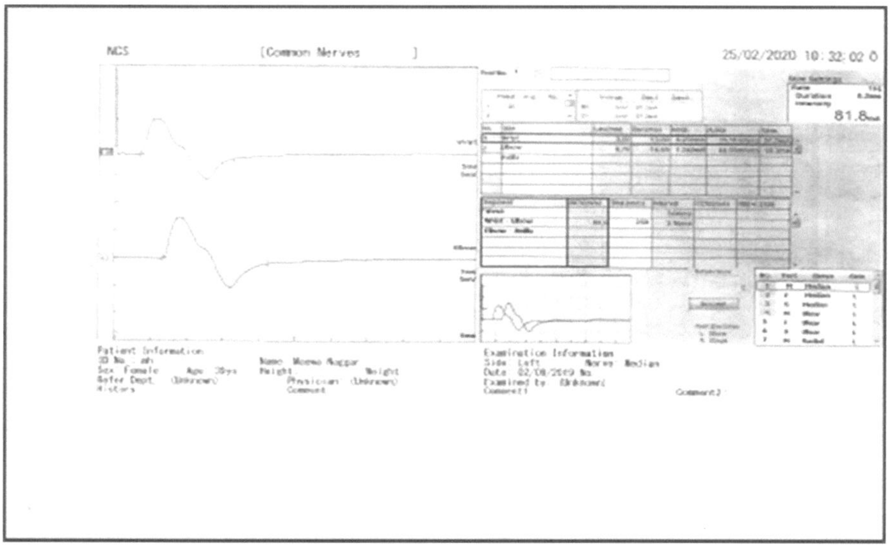

Figure 14.
A plot of the stimulating signals and evoked signals from a stimulation device as a record for the nerve impulses from the arm and leg of a patient [A hospital print].

membranes ΔT which is converted into electric potential when crossing sodium-potassium junctions at the sides of the membranes by Seebeck Effect [55]. Hence, the neural cell performs the function of thermoelectric generator to convert part of the heat produced by the metabolic operation of the neural cells into electric signals by converting its thermal potential ΔT, measured across the membranes of the neural cells, into electric potential ΔV [69]. However, the potential of the nerve impulse is magnified when crossing legs of the cell membrane formed of successive sodium - potassium junctions or legs by accumulating the gained electric potential [69].

As a conclusion, the proper understanding of the electric charges as EM waves of electric potential may end the confusion of the neurologists regarding creation and motion of unknown action potential. As has been previously analyzed, **Figure 8** shows the stimulating signal in electric field E - Entropy (s) plane where the entropy flow rate "$\dot{S}$" measures the capacity of the neural tissues to allow the flow a definite electric power per unit potential.

At the right corner of the card in **Figure 14**, it is seen the evoked signals as responses to the stimulating signals in **Figure 8**. It is shown as a nerve impulse sent in the form of an electric signal or charge like the shape of the stimulus charge in **Figure 8**. It is also an EM wave whose abscissa is the measured growth of entropy, measured by an ammeter in nano Joule/volt, and whose ordinate is the potential of the charge in milli Volts. The fast response of the stimulating signal by an evoked signal shows the speed of flow of both signals as EM waves and the impossibility to be a flow of electrons that should not flow through organic tissues. The power delivered by the flow by such nerve impulse can be calculated also by the same procedure followed to calculate the energy and power of the stimulus charge. It is a fundamental comment to neurologists to understand how the plot of the stimulating charge depends on the specification of the source of impulse and the properties of the tissues of the neural cell defined by the entropy, which is ignored by the neurologists, that has a great value in specifying the health of the neurons. Such data may also help in diagnosis of the defects in the neural system.

8. Conclusions

a. The heat, the electric charge, and the magnetic flux are forms of energy that have a common origin whose SI unit is the "Joule."

b. The electric charges are EM waves which are derived by their electric potential, and the magnetic flux is EM waves which is derived by its magnetic potential as the thermal radiation which is derived by the force of its thermal potential. All potentials have a common reference scale, the thermodynamic temperature scale, and one common unit, the "Volt." There are clear dependence of the electric and magnetic potential scales on the thermodynamic temperature scale.

c. The ammeter's reading has the SI unit "Watt/Volt" that measures the capacity of the electric field to pass a definite rate of flow of electric energy, i.e., electric power, by the force of a unit electric potential power. This capacity measures also the rate of growth of the entropy of such electric field per unit electric potential.

d. The Ampere is a meaningless unit of a wrongly defined identity that should be dropped from the base units of the SI system of units as a necessary modification of the system to remove its confusing redundancies. The Kelvin will be the base unit for all potentials but its unit its will be the "Volt."

e. The magnetic induction is a process of exchange the electric potential of the input electric charges by magnetic potential and the opposite exchange can be done.

f. The thermoelectric and photoelectric effects convert the thermal potential of the incident electromagnetic radiation into electric potential when crossing junctions of dissimilar materials due to differences in their energy bandgaps by the Seebeck effect.

g. Casting the Maxwell's wave equation into an energy frame of reference that replaces the time by entropy as an arrow of time visualize the electric charges and magnetic flux as EM waves that have electric or Magnetic potentials.

h. The energy flow as EM waves is quantized into wave-packets or energy quanta each of energy: $1.423\ \lambda_s (E^0 + H^0)$ Joule/cycle where λ_s is the wavelength along the entropy ordinate in the energy frame of the modified Maxwell's equation. Such quanta are independent on the frequency of the wave. This statement violates Planck's statement which wrongly assumes such quanta is proportional to the frequency as any multiple of frequency defines rates of flow and does not define a quantity of energy.

i. Einstein's assumption that the claimed quanta of Planck may be particulated and hits the electron out of its orbit is a false statement according to the negligible momentum of the energy-quanta if compared to the momentum of a moving electron. This assumption fails to explain emission of electrons from organic PV cells or in multijunction cells and how the PV efficiency exceeds the limit determined by who followed Einstein's assumptions. As a more plausible explanation, the thermal potential of the incident thermal radiation is converted when crossing junctions of dissimilar materials of different bandgaps into electric potential according to Seebeck equation ($\Delta V = \propto\_AB \Delta T$).

j. Neurologists saw and defined the nerve impulses as electric signals, but they wrongly interpreted such signals as a moving imaginary-action potential. They believed a definition of the electric signal should involve flow of electrons and denying their observation of the possibility of flow of electric current through a patch of cell membranes. Their solution is to believe the electric charges are not electrons as the electrons are particles, but they are EM waves that have electric potential which they call Action Potential." So, the nerve impulses are electric signals, as they define, and such electric signals are electric charges which have "Action Potential." [This approach will answer, as mentioned in previous research, all their inquiries concerning the generation of the nerve impulse as electric signal of action potential, correct interpretation of the stimulating signal and evoked signals, and the entropy as a property that determines the health of elements of the neural systems.

Acknowledgements

The author thanks Allah for his guide in writing this chapter I should also thank my Dear Son: Prof. Dr. Mohamed Salama, Dean of the Engineering College, Beni Suef University in Egypt, for his sincere belief of my ideas that encourages me.

Author details

Salama AbdelHady
Faculty of Energy Engineering, Aswan University, Egypt

*Address all correspondence to: salama_abdelhady@aswu.edu.eg

References

[1] Jewett J, Serway A. Physics for Scientists and Engineers with Modern Physics. 7th ed. Brooks: Thomson; 2008

[2] AbdelHady S. A fundamental equation of thermodynamics that embraces electrical and magnetic potentials. Journal of Electromagnetic Analysis & Applications. 2010;**2**: 162-166. DOI: 10.4236/jemaa.2010.23023

[3] Stosur M, Dawidowski P, Szewczyk M. Modeling, and simulation of thermal behavior within typical GIS busduct using ATP-EMTP program. In: European EMTP-ATP Conference. Cagliari, Italy: GMSARN International Journal; 2014. DOI: 10.13140/2.1.1783.5526

[4] Faraday M. Experimental researches in electricity, nineteenth series. Philosophical Transactions. Royal Society of London. 1846;**136**:1-20. DOI: 10.1098/rstl.1846.0001

[5] Baigrie B. Electricity and Magnetism, A Historical Perspective. NY, USA: Green Wood Press; 2007. ISSN 1559-5374

[6] AbdelHady S. Thermodynamics: Fundamentals and its Application in Science, Auris Reference. 1st ed. London (UK): An International Text Book In Science; 2017. ISBN-10: 1788020626

[7] AbdelHady S, MS AH. An entropy approach to the natures of the electric charge and magnetic flux. Journal of Electromagnetic Analysis & Applications. 2015;**7**:265-275. DOI: 10.4236/jemaa.2015.711028

[8] Laurendeau NM. Statistical Thermodynamics. 1st. ed. England, UK: Cambridge university press; 2015. DOI: doi.org/10.1017/CBO9780511815928

[9] AbdelHady S. Review of thermodynamics of systems that embrace transfer of electric and magnetic energies. Journal of Physical Science and Application. 2018;**8**:1-12. DOI: 10.17265/2159-5348/2018.01.001

[10] Naterer GF, Camberos JA. Entropy and the second law, fluid flow and heat transfer simulation. Journal of Thermos-Physics and Heat Transfer. 2003;**17**: 360-371. DOI: org/10.2514/2.6777

[11] AbdelHady S, Chi-Ho C. Advanced Thermodynamics Engineering. 1st ed. New York: Scitus Academics; 2019 ISBN: 9781681179803

[12] Sluka KA, Walsh D. Transcutaneous electrical nerve stimulation: Basic science mechanisms and clinical effectiveness. The Journal of Pain. 2003; **4**:109-121. DOI: 10.1054/jpai.2003.434

[13] AbdelHady S. A thermodynamic analysis of energy flow in optical fiber communication systems. Applied Physics Research. 2012;**4**:22-29. DOI: 10.5539/apr.v4n3p22

[14] Franklin A. Millikan's measurement of Planck's constant. The European Physical Journal. 2013;**38**:573-594. DOI: org/10.1140/epjh/e2013-40021-3

[15] Weiling L, Shantung TU. Recent developments of thermoelectric power generation. Chinese Science, Bulletin. 2004;**49**:1212-1219. DOI: doi.org/ 10.1360/04we0037

[16] AbdelHady S. Innovative solutions of unsolved problems and misconceptions in physics. International Journal of Innovative Research in Science, Engineering and Technology. 2017;**6**:9. ISSN (Print): 2347-6710[17]

[17] Borelius G, Keesom WH, Johannson CH, Linde JO. Establishment

of an absolute scale for the Thermo-electric force. Proceedings of the Royal Academy of Sciences at Amsterdam. 1932;**35**:10

[18] Phan M, Yu S. Review of the Magnetocaloric effect in manganite materials. Journal of Magnetism and Magnetic Materials. 2007;**308**:325-340. DOI: 10.1016/j.jmmm.2006.07.025

[19] AbdelHady S. Innovative understanding of the duality confusion, the photovoltaic and the Magnetocaloric effects. Ain Shams Engineering Journal, Engineering Physics and Mathematics. 2018;**9**:2283-2289. http://creativecommons.org/licenses/by-nc-nd/4.0\

[20] Brown GJ. Magnetic heat pumping near room temperature. Journal of Applied Physics. 1976;**47**:8. DOI: doi.org/10.1063/1.323176

[21] Caron L, Doan N, B, Ranno L. On entropy change measurements around first order phase transitions in caloric materials. Journal of Physics. Condensed Matter. 2017;**29**(7):3673–3680. DOI: 10.1088/1361-648X/aa50d1

[22] Timothy HB. Faraday induction and the current carriers in a circuit. American Journal of Physics. 2015;**83**: 263-270. DOI: doi.org/10.1119/1.4901191

[23] Fujimoto M. Physics of Classical Electromagnetism. 1st ed. Library of Congress Control Number: 2007921094. Berlin, Heidelberg, New York: Springer link; 2007

[24] Ryan DT. Toward a cognitive-historical understanding of Michael Faraday's research. Perspectives on Science. 2006;**14**:1-6. DOI: org/10.1162/posc.2006.14.1.1

[25] Clerk-Maxwell J. A dynamical theory of the electromagnetic field. Philosophical transactions of the Royal Society of London. 1865;**155**:459-512

[26] Brown J. Electromagnetic wave theory. In: Proceedings of a Symposium Held at Delft. 1st ed. The Netherlands: Elsevier; 1967. eBook ISBN: 9781483185927

[27] Myrvold WC. Statistical mechanics and thermodynamics: A Maxwellian view. Studies in History and Philosophy of Modern Physics. 2011;**42**:4. DOI: doi.10.1016/j.shpsb.2011.07.001

[28] Riffat SB, Ma X. Thermoelectrics: A review of present and potential applications. Applied Thermal Engineering. 2003;**23**:913-935. DOI: doi.org/10.1016/S1359-4311(03)00012-7

[29] Nikulov A. The law of entropy increase and the Meissner effect. Entropy. 2022;**24**:1-23. DOI: 10.3390/e24010083

[30] Borg X. Unified theory foundations-the ST system of units. Blaze Labs Research. 2022;**24**(3):1-23. DOI: blazelabs.com/f-u-suconv.asp

[31] AbdelHady S. An advanced review of thermodynamics of electromagnetism. International Journal of Research studies in Science, Engineering and Technology. 2015;**4**(6): 1-12. DOI: 10.17265/2159-5348/2018.01.001

[32] AbdelHady S. Comments on Einstein's explanation of electrons, photons, and the photo-electric. Effect. Applied Physics Research. 2011;**3**: 230-240. DOI: 10.5539/apr.v3n2p230 30

[33] Shahin AZ, Ismaila KG, Yibas BS, Al-Sharafi A. A review on the performance of photovoltaic/thermoelectric hybrid generators. International Journal of

Energy Research. 2022;**1**:3366-3394. DOI: org/10.1002/er.5139

[34] Heisenberg W. Über den anschaulichen Inhalt der quantentheoretischen Kinematik und Mechanik. Z. für Physics. 1927;**43**: 172-198. DOI: 10.1007/BF01397280

[35] Landi GT, Paternostro P. Irreversible entropy production: From classical to quantum. Reviews of Modern Physics. 2021;**93**:3. DOI: doi.org/10.1103/ RevModPhys.93.035008

[36] AbdelHady S. A substitute to the quantum theory by an entropy approach. In: 7th International Conference on Mathematics and Engineering Physics (ICMEP-7). Cairo, May: International Publisher; 2014. DOI: 10.21608/ icmep.2014.29734

[37] Norton J. The logical inconsistency of the old quantum theory of black body radiation. Philosophy of Science. 1987; **54**:327-350. DOI: jstor.org/stable/187578

[38] AbdelHady S. An entropy-approach to the duality property. Journal of Electromagnetic Analysis & Applications. 2011;**3**:220-227. DOI: 10.4236/jemaa.2011.36036 P

[39] Jaynes ET. Gibbs vs Boltzmann Entropies. American Journal of Physics. 1965;**33**:391-398. DOI: org/10.1119/ 1.1971557

[40] AbdelHady S. An entropy approach to a practical limit of the Efficiency of developed and multijunction solar cells. Journal of Electromagnetic Analysis & Application. 2014;**6**:383-390. DOI: 10.4236/jemaa.2014.613039

[41] Abu-Bakr M. On dimensional analysis, redundancy in set of fundamental quantities and proposal of a new set. General Physics. 2014;**6**(13): 383-390. DOI: researchgate.net/ publication/1771012

[42] Yunus AC, Michael AB. Thermodynamics: An Engineering Approach. 8 th. ed. NY, USA: McGraw-Hill; 2006. ASIN: B099SFL4P4

[43] Steiner R. History and progress on accurate measurements of the Planck constant. Reports on Progress in Physics. 2013;**6**:2. DOI: 10.1088/0034-4885/76/1/ 016101

[44] AbdelHady S. An approach to a universal system of units. Journal of Electromagnetic Analysis & Applications. 2010;**2**:549-556. DOI: 10.4236/jemaa.2010.2907/

[45] AbdelHady S. Comments concerning measurements and equations in electromagnetism. Journal of Electromagnetic Analysis & Applications 2010;2:677-678. DOI: Sci RP.org/journal/je maa\

[46] AbdelHady S. An advanced review of thermodynamics of electromagnetism. International Journal of Research studies in Science, Engineering and Technology. 2015;**3**(6):10

[47] AbdelHady S. Comments on Einstein's explanation of electrons, photons, and the photo-electric effect. Applied Physics Research. 2011;**3**: 230-240. DOI: 10.5539/apr.v3n2p230/

[48] AbdelHady S. Entropy analysis of the duality property. J. Electromagnetic Analysis & Applications. 2011;**2**:217-219. DOI: 10.4236/jemaa.2011.36036

[49] Nolas GS, Sharp J, Goldsmid J. Thermoelectrics: Basic Principles and New Materials Developments. Vol. 45. Beijing, China: Springer; 2001. pp. 432-440 sn. pub/extras

[50] Zhao KL, Chen KH. Electromagnetism. 3rd ed. Beijing: Higher Education Press; 2010

[51] AbdelHady S. Advanced physics of thermoelectric generators and photovoltaic cells. American Journal of Physics. 2018;**33**:391-398. DOI: org/10.11648/j.ajpa.20180605.14

[52] Christophe G et al. Thermodynamics of thermoelectric phenomena and applications. Entropy. 2011;**13**:1481-1517. DOI: 10.3390/e13081481

[53] Blum I et al. Effect of electrical conduction on the electron emission properties of diamond needles. New Journal of Physics. 2020;**22**: 083044

[54] Saqib R et al. Thermoelectrics by a solvent-free drawing method of all carbon-based materials. ACS Omega. 2021;**6**:5019-5026. DOI: org/10.1021/acsomega.0c062

[55] Herwaarden AW, Sarro PM. Elsevier. Thermal Sensors Based on The Seebeck Effect, Sensors and Actuators. 1986;**10**:321-346

[56] Haddad D et al. Measurement of the Planck constant at the National Institute of Standards and Technology from 2015 to 2017. Metrologia. 2017;**54**:633-641. DOI: org/10.1088/1681-7575/aa7bf2

[57] Falk G, Herrman F, Schmid GB. Energy forms or energy carriers? American Journal of Physics. 1983;**51**: 1074-1077. DOI: 10.1119/1.13340

[58] AbdelHady S, AbdelHady MS. Analyzing the duality concept through an entropy approach. Proceedings of ASAT-14-188-TH. 2011

[59] Krause R et al. Wafer bonded 4-junction GaInP/GaAs//GaInAsP/GaInAs concentrator solar cells. AIP Conference Proceedings. 2014;**1616**:45-49

[60] Burnett B. The Basic Physics and Design of III-V Multijunction Solar Cells. NY, USA: National Renewable Energy Laboratory, Golden, Co; 2002. Available online: http://www.uotechnology.edu.iq/eretc/books/NRELokok.p

[61] High-Efficiency YN. Multi-Junction Solar Cells: Current Status and Future Potential. Ottawa: University of Ottawa; 2007 http://sunlab.site.uottawa.ca/pdf/whitepapers/HiEfficMjSc-CurrStatus&FuturePotential.pd

[62] Shockley W, Queisser HJ. Detailed balance limit of Efficiency of p-n junction solar cells. Journal of Applied Physics. 1961;**32**:510-519. DOI: 10.1063/1.173603

[63] AbdelHady S. An entropy approach to Tesla's discovery of wireless power transmission. Journal of Electromagnetic Analysis and Applications. 2013;**5**: 157-161. DOI: 10.4236/jemaa.2013.5402

[64] Zlatanoy N. Lasers and laser applications. IEEE Computer Society. 2002:inproceedings{2002TheBP}. DOI: 10.13140/RG.2.2.26771.89123

[65] Yamaguchi M, Dimroth F, Geisz J, Ekins-Daukes NE. Multi-Junction Solar Cells Paving the Way for Super High-Efficiency, DOE pages, OSTI.GOV Journal Article; 2021:10. DOI: 10.1063/5.0048653

[66] Brown T. P, Eppley K, High-power, high-energy, and high-efficiency, phase-locked magnetron studies. IEEE Physics. 1961;**32**:510-519. DOI: 10.1063/1.1736034

[67] Benjamin D et al. Thinking about the nerve impulse: A critical analysis of the electricity centered conception of nerve

excitability. Progress in Neurobiology. 2018;**169**:172-185. DOI: 10.1016/j.pneurobio.2018.06.009

[68] Andersen SSL, Jackson AD, Heimberg T. Towards a thermodynamic theory of nerve pulse propagation. Progress in Neurobiology. 2009;**88**: 104-113. DOI: 10.1016/j.pneurobio.2009.03.002

[69] AbdelHady S. Innovative definition of nature of the nerve impulses. Ain Shams Engineering Journal, Engineering Physics and Mathematics, Production and hosting by Elsevier. 2019;**11**:2. DOI: 10.1016/j.asej

[70] Squire L et al. Fundamentals of Neurosciences. 3rd ed. Washington, D.C., USA: Elsevier; 2013

[71] Raichel ME. Feature review: Two views of brain function. Trends in Cognitive sciences. 2010;**14**:180-190. DOI: doi.org/10.1016/j.tics.2010.01.008

Chapter 3

The Electric Fields of Lightning Clouds in Atmospheres of Different Properties

Rafael Zamorano Ulloa

Abstract

In this work, we concentrate on the electric fields produced by any distribution or structure of electric charges in clouds, these quantities are, for the most part, unknowns, and still we extract valuable information from their vectorial properties and working within close distances, especially on the structures that produce pronounced sinks/sources with large divergences of $\vec{E}$ and focusing on observation points close to them. We introduce the concept of "at the verge of discharge electric field," $\vec{E_\xi}$ produced by two specific charge configurations. We detail the role the dielectric character of the cloud plays in allowing/not allowing discharge. We profile the conditions to be met by $\vec{E_\xi}$ to appear and to be sustained for at least a few milliseconds. One objective of this work is to calculate some relevant quantities for a typical thundercloud modeled just at the verge of discharging. A simple electrified cloud model containing many of the features widely reported in the literature is constructed and described with some detail. The charge structures possible are delineated and the electric fields produced described. A few basic calculations are carried out for a thought experiment, moving charges around, we calculate some velocities and accelerations and analyze what we learn from it.

Keywords: lightning, electrified clouds, charging of clouds, at the verge of lightning electric field, charge structure, sprites, gigantic jets, blue jets

1. Introduction

Clouds as cotton fields taking a diversity of colors, or ragged white milk-like spreads, or tall, very tall cloud compounds attract the sight and inspiration of people everywhere and produce awe in some of us, as shown in **Figure 1A**. But then, electrified clouds and then their lightnings, thunders, and copious precipitations have scared and fascinated humans since immemorial times [1, 2]. Beyond the gods associated with them in different cultures [3, 4], these thunderclouds sustain a great amount of energy, in the form of hydrodynamic energy and electromagnetic energy. They also carry great amounts of water [5, 6], and they also carry and concentrate on the electric charge, moving it and distributing it in different arrangements until they can carry no

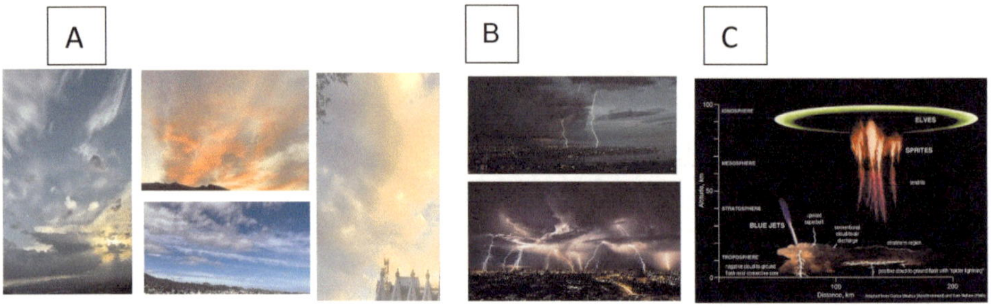

Figure 1.
Diverse cloud formations and colors and sizes and lightnings: (A) clouds as cotton fields taking a diversity of colors, or ragged white milk-like spreads, or tall, very tall cloud compounds; (B) very extensive cloud formations lightning above cities and the sea; and (C) an assortment of different phenomena directly produced by discharging electrified clouds.

more, as show in **Figure 1B**. The charging and charging processes cannot go forever, since the electric energy, proportional to the total charge accumulated [7–9] would tend to infinity, then at some point the cloud system becomes electromagnetically unstable, I even say quite unstable. Lightning is fundamentally a process of electrical discharge of an electrified cloud, a thundercloud, but it is not the only one, gigantic jets launched into the sky is another one. Lightning can be very rapid and violent and quite sonorous and luminous [10–12]. It is now well established [10–15] that these cloud electric discharges go down to ground, or mountains, or tall buildings, metallic structures, trees, animals, and people [10–15], and they also go up the sky and can terminate just above the cloud in the sky [16–18] or go higher and reach the ionosphere (around 90 km above see level) and beyond [18–20]. These energetic (thousands of joules) discharges frequently are accompained or followed or preceded by light, UV rays, gamma rays, bremsstrahlung [21–24], and/or X-rays. **Figure 1** illustrates some cloud formations and some of these processes.

Lightnings are also preceded by torrents of classically accelerated electrons, protons, small ions, even relativistically moving electrons, gamma rays, and UV-vis [20–24]. These discharges happen frequently from cloud to ground, from cloud to cloud, and from a region, A, to a region, B, within the same electrified cloud. It has been estimated that about 9,000,000 lightning discharges happen worldwide every day [25]. These everyday cloud electromagnetic phenomena bring with them, even more electromagnetic phenomena, including light at different wavelengths, gamma ray and X-ray radiation, elementary particles as electrons, neutrons, and positrons, radiofrequency signals and sound [26, 27], sprites, elves, and glows. All this happens inside the clouds, around the clouds, and kilometers away from the clouds, as shown in **Figure 1**. In order to understand the physics of all this, it is necessary to get to know more in detail what a cloud is, its basic charged and neutral components, its turbulent internal motions, and the different charging mechanisms and discharging mechanisms. It is necessary to know the different charge structures and the electric fields, $\vec{E}$, with their divergences, $\text{Div}\cdot\vec{E} = \rho_{TOT}$, produced by these tremendous moving electrified cloud bodies. Some great amount of electric charge, positive and negative, remains quasistatic for short periods of time (of the order of milliseconds), located in specific regions at different altitudes [25–27] as represented in **Figures 1C** and **2**. Other charged elements are much more mobile and eventually can confirm electric

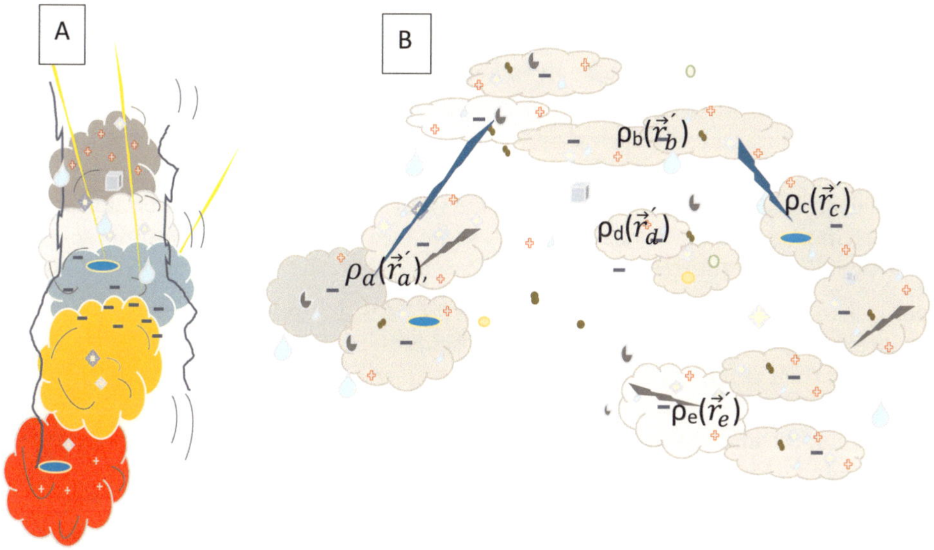

Figure 2.
Electrified clouds and some of their constituents: (A) a tall, moving, multicolor, electrified cloud bathed by cosmic radiation (blue tortous rays) and solar rays (yellow ray) with charge distributions located at different altitudes. Water droplets, ice, graupel, cations, and anions are some of the hydrometeors in it; and (B) a conglomerate of charged clouds with intra- ( ) and intercloud ( ) discharging and some other cloud components shown: Water droplet ( ), graupel ( ), ice ( ), hail ( ), molecular ion ( ), snow ( ), and pollutant particle ( ). The density of charge within each cloud is symbolically represented as $\rho_i(\vec{r}_i')$ with i = a, b, c, d, …, and conducting patches are shown as ( ).

currents within conductive patches that form and disappear in micro- or milliseconds [28] as shown in **Figures 1B**, **C** and **2**.

At the level of individual charged particles, it is necessary to know the pathways, velocities, accelerations, and directions these charges follow during the discharges. We, now, know that all types of charges, prominently electrons, can be involved in the cloud discharges [29–32]. Charges in clouds include, but are not limited to, electrons, e^-, protons, p^+, water droplets, $w^\pm$, small and large ice cristals, $ic^\pm$, graupel (ice covered by water), $g^\pm$, hail, $h^\pm$, all kinds of typical atmospheric cations, ci^+, and anions, ai^-, and polarized atoms and molecules, and continuous and permanent elements of cosmic rays [29–32]. Several of the above-mentioned characteristics are illustrated in **Figure 2**. Part of the fundamental electromagnetic problem that an electrified cloud poses (one that can become a thunder cloud) is to know, to determine its charge structure, how it is distributed in cloud space and time, and the electric field(s) it produces. As Griffiths puts it [33] (more in general, not particularized to clouds): The fundamental problem a theory of electromagnetism hopes to solve is this: I hold a bunch of charges here, ρ_1 or Q_{source}, (and may be shaken around)—what happens to some other charge, q, over there? The classical solution takes the form of a field theory. We say that the space around an electric charge is permeated by electric and magnetic fields (the electromagnetic "odor," as it were of the charge). The second charge in the presence of these fields experiences a force (Coulomb force); the fields then transmit the influence from one charge to the other —they mediate the interaction. Refer to the charges in **Figure 2**. And we add: The knowledge of such a bunch of charges, $\rho_1\left(\vec{r}',t\right)$ or Q_{source} "here" and the fields and forces they produce "there" allow us to determine the trajectories, velocities, and energies, that the other charge(s) q ($\vec{r}$, t) will acquire. From this, we can calculate currents, velocities, accelerations, and radiation fields produced but only when the dielectric and conduction properties of the medium allow [7–9, 33]. We want to apply this to electrified clouds. In this chapter, we want to pose this fundamental problem for electrified clouds and contribute with some basic calculations toward its solution. In particular, we want to be as explicit as possible with the source charges (distributions), their accumulation "here," the electric forces, $\vec{F}$, and electric fields, $\vec{E}$, they produce, and how these forces and electric fields affect (many) other charges, q, "over there" and their discharge and accompanying glows. A cloud that has accumulated a great amount of charge can discharge it toward other clouds, ground, mountains, metallic objects, and ionosphere. When the source charges in electrified clouds are just too much to be held, they discharge toward other charged conducting objects. This can be accompanied by light and sound, and then we have lightning. Lightning continues to be a mystery to science [1, 2, 20, 25]. It is not clear at all how it is initiated, nor how are reached the values of electric field required for lightning to initiate and the many effects lightning produce above and below the cloud. The charging processes, the discharging processes, and the electric fields involved are at the center of the unsolved questions [10–12, 25]. We should stress that for more than a century now, a great number of researchers have worked on these, and their insights, measurements, contributions, and acquired knowledge are vast [3–32]. We, certainly, drink from this vast reservoir of knowledge. Yet, it is recognized that these fundamental problems continue to be open, and they figure at the top of a list of 10 basic unanswered questions [25] in the physics of electrified clouds research.

Here, we construct a simple electrified cloud model containing many of the features widely reported in the literature, and we describe it in some detail. Our description is as complete and detailed as possible, then we perform a few basic calculations on it and perform a thought experiment, moving charges around inside this model cloud, and we calculate some quantities and analyze what we learn from it. But first we want to mention, in some detail, that electric charges are really very common in this world and how charges are produced in different materials and, of course, in clouds that become heavily electrified.

2. Charges are everywhere in the world and the universe

Electric charge is a fundamental feature of a great number of material structures, small, large, mesoscopic, microscopic, of this world, and indeed of the universe [34–38]. Some examples are water droplets, little and large ice crystals, cell membranes, cut fingernails, ashes from volcanoes, pollutant particles, the earth, the ionosphere, clouds, balloons, atoms, electrons, protons, amber, fur, glass, straw, and small pieces of wood. Charged, electrified, and objects have been known for millenia. The Greeks were the first to describe some of them, like amber and fur, and they also discovered how to electrically charge them, by simple friction! By simply friction one against the other, they produced the effect of electrostatics. They also discovered that there are two types of "electricity" that later were named "positive" and "negative" and that positive-positive, $+q \leftrightarrow +q$, (negative-negative, $-q \leftrightarrow -q$) charged bodies repeal and that positive-negative, $+q \leftrightharpoons -q$, and negative-positive, $-q \leftrightharpoons +q$ bodies attract [34–37].

Natural, strongly electrical phenomena, out of man's direct manipulation, as clouds and thunderclouds and lightning were also known from antiquity. In several cultures, there was the god of thunder [39, 40]. It was also known that electrified bodies can be discharged and that charge can "move" and migrate can be transfered from one place to another, constituting the electric current, $I = \Delta q/\Delta t$. Latter on, glass, water flowing, pieces of cloths, rubber, rugs, metal knobs, comb(s), metal cages of washing machines and of refrigerators, sea water, wet floor, our hands, our body, pollutant particles, etc. [41, 42] were, then, by experimenting, included in that list of chargeable systems. When a lot of charge is accumulated by any of these ways, irrespective of the particular mechanism, we take the total charge, $Q = \sum_i^n qi$ when charge is an accumulation of discrete elements and $Q = \int_q dQ$ when modeled as a continuous distribution of charge.

It is also very convenient to use its volumetric density, $\rho = Q/V = (1/V)\sum_i^n qi$ or $\rho = (1/V)\int_q dQ$ and their surface density, $\sigma = Q/S = (1/S)\sum_i^n qi$ or $\sigma = (1/S)\int_q dQ$ and their linear density, $\lambda = Q/\ell = (1/\ell)\sum_i^n qi$ or $\lambda = (1/\ell)\int_q dQ$.

With the execution of specific experiments and development of technology, more charged objects were known like the popular electrostatic bottles, as shown in **Figure 3** [43, 44]. It is very impressive how they can be charged and how they can be discharged in just a small fraction of a second, electric current, I, traveling through the air and producing a spark, light, and sound, and then it came the "famous" Benjamin Franklin's kite, as shown in **Figure 3E**, [44, 45], and then the whole of the early history of the research on charged clouds and their ways of discharging [46–48] mainly into ground or to a tree, or to a church bell (metallic), or on another nearby cloud. During discharge, electric current will travel from the cloud to ground and from ground to cloud, or from cloud to cloud producing, not always, light and sound. Several other thundercloud and lightning events were observed, reported, and "described" many years ago, and some of them are bolts, sprites, glows, blue jets, and gigantic jets [1, 2, 10–16]. In more recent times, novel discharge mechanisms and electromagnetic emissions have been observed and measured from ground, planes, aerostatic globes, and satellites [10–16]. From the more physical sciences view, the very same Coulomb Balance of 1785 (modified from the Cavendish balance used to measure, quantitate, the force between charged bodies) allowed Charles Coulomb to find $\vec{F} = Kq_1q_2\hat{r}/r^2$. This universal law indicates that the force between two charged bodies is inversely proportional to the square of their

Figure 3.
Charging and discharging processes in a variety of systems presented from present-day machines to the oldest ones: (A) a laboratory-controlled discharge device, 1990s; (B) a Van de Graaff electrostatic generator, 1910; (C) Volta's battery presentation to Napoleon, 1810; (D) a basic electroscope, 1777; (E) a painting showing Benjamin Franklin's kite experiment,~ 175; and (F) Amber friction with fur, a pictorial representation of Tales de Mileto discovery~ 600 BC and a plastic comb attracting small pieces of paper, (generalizes the amber-fur electrostatics).

distance and to the product of their charges and is directed along the line that joints them. It also includes the experimental fact that charges of the same sign repeal and charges of opposite sign attract. To do the experiments, Coulomb had to charge, electrify, the different bodies he used, operation that he did, mainly, by friction. The electroscope was invented, and it made possible some more quantifications [49]. Several of these developments are pictured in **Figure 3**.

Later, the Van de Graaff electrostatic machine [50] was invented to specifically charge conducting bodies like spherical shells, and so on. More developments in the knowledge of electrical charge, its motion, and some applications constitute the invention of the battery [51] and its whole development. Faraday introduced the concept of electric field that is defined as $\vec{E} = \lim(q\text{-} > 0)\ \vec{F}/q$, taking the Coulomb force just mentioned, and we have $\vec{E} = \lim(q\text{-} > 0)\ [KQ\hat{r}/r^2]$. Conductors and non-conductors were then defined [7–9, 52], and microscopic mobile units of charge, electrons, were suspected to form a part of the intimate nature of matter [53]. Now, we know that the atom is full of charge specifically distributed, the nucleus is positive charged, +q, and the "revolving" electrons are negatively charged, −q. An atom is electrically neutral, but charged species, ions, are very common, and some examples are O_2^-, H^+, K^+, Cl^-, and so on. The whole material universe is constructed from electrons, protons, nuclei with protons and neutrons inside, atoms, then molecules, the chemical bond, minerals, rocks, hydrogen species, oxygen species, organic, molecules, polymers, and so on, and they are in all places, corners of the universe in the form of gasses, mesoscopic and galactic plasmas, rocky planets, moons, and atmospheres. Physical processes as friction, collisions, scatterings, breakage, fractionation, and intense electric fields are universal phenomena and separate charges from their parent matter; hence, free charge can appear potentially in every corner of the universe, generating diverging and/or rotational electric fields. Stellar rocky bodies with some kind of atmosphere can concentrate on the electric charge in somewhat stable arrangements, and free electrons generated by cosmic radiation can be accelerated in those atmospheres. Coming back to earth, we see below that such acceleration of electrons already takes place in electrified earth-bound clouds.

3. Charges can be produced in many different ways

To understand better the charging processes in clouds, we mention, first, how charges can be produced in many different ways: the most common ones are by friction, fractionation, breakage (breaking apart polymers and small pieces of paper), and others include thermoionization, just heating of a conducting cable with the passage of current, heating a gas, air, application of an intense electric field, UV-vis excitation/photoionization (this is one of the reasons we have life in this planet in the first place), photoionization in the atmosphere [54], electrolysis [55], battery and fuel cell mechanisms [56], xerox copying, ionizing radiation (NO, H, O_2, H_2O, organic molecules, and nucleic basis are very sensitive to ionizing radiation), cosmic radiation, collisions, and near collisions of many kinds: electron-atom, electron-molecule, atom-atom, atom-molecule, elementary particle-atom/molecule, scattering of gamma and/or X-rays, and UV rays scattered by atoms or molecules [53], and so forth.

In neutral gasses charging mechanisms, charge separation include elevated temperatures promoting more atom/molecule frenetic motion, more and more energetic collisions of atoms/molecules, molecular friction (then the asymmetric breaking of them), very strong turbulent fluxes (turbulent winds in diverse directions) promoting, again, energetic collisions where linear and angular momentum, energy, and charge are exchanged or transferred, and exposure to energetic gamma rays, X-rays, and/or cosmic radiation, application of a high-voltage difference, ΔV, or equivalently a strong electric field, $\vec{E} = -\nabla V$(very strong, of the order of hundreds of thousand Volts/mt), could be enough to break molecules and to break polarized molecules, or at least rip them from one electron, leaving behind a cationic, heavy, species. Discharged gasses of detectors and lamps are good examples of these systems [57, 58]. If the gas is basically dielectric, a moderate to high electric field will only polarize the molecules of the system producing a total macroscopic polarization vector $\vec{P}$ that points in opposite direction from $\vec{E}$. More charging mechanisms are man-made accelerators, or nuclear plants with heavy radiation, and cosmic rays, and naturally radioactive elements present in the medium can produce very energetic electrons or positrons or neutros from any of these bombardment machines. These accelerated species, mainly, but not exclusively, electrons, will collide with and/or scatter from bigger molecules of the medium and produce more ionized species and more fast-moving electrons, and they have the possibility of producing avalanches of electrons and charged species [59, 60]. When the gas is mostly isolated and calm, the less the above charge-producing processes take place and the gas is mostly neutral.

In plasmas-charging mechanisms, charge separation include the same ones present in a gaseous system, but now there is an important manner conductivity, σ. This conductivity is due to permanently present, replenishing, anions (−, include electrons, O_2^-, other oxygen and hydrogen species) and cations (+, include positively charged species, molecules, small ice crystals, positrons, protons, etc.). They do not need to be permanent species, and they could be transient with an active production mechanism, for example, cosmic radiation as in the ionosphere. This conduction property provides the plasma with very particular electromagnetic properties that nonconducting gases do not have. In fact, the Maxwell equations valid for a plasma are as follows:

$$\mathrm{Div}.\vec{E} = \rho_{tot}/\epsilon_o \tag{1}$$

$$\mathrm{Rot} \times \vec{E} = -\frac{\partial \vec{B}}{\partial t} \tag{2}$$

$$\mathrm{Div}.\vec{B} = 0 \tag{3}$$

$$\mathrm{Rot} \times \vec{H} = \mu_o \left(\vec{J}f + \frac{\partial \vec{P}}{\partial t} + \mathrm{Rot} \times \vec{M} + \frac{\epsilon \partial \vec{E}}{\partial t} \right) \tag{4}$$

where ϵ_o is the free space electrical permitivity, ϵ is the medium permitivity that could easily be larger than ϵ_o by a factor of 10, 20, or even more depending on the content of polarizable molecules in the medium. And the conductivity is given by σ [35, 58].

4. In electrified clouds, charging mechanisms are complex

A cloud is a very complex hydrodynamic and electrodynamic system, and it is heterogeneously conformed. A wealth of phenomena actually occurs inside and on the surroundings of a simple or complex cloud. Just to mention some: they may aggregate looking as calmly cotton fields, but then again they can become thunderous and vertiginously electric active releasing thousands of joules of energy in their lightnings, they can show an halo-glow above them, they can produce stripes and Saint Elmo's fire, their internal energy can accelerate electrons to relativistic velocities and runaway electrons, and then cascades of even more electrons in relativistic regimes, they can produce electromagnetic radiation fields, like bremsstrahlung, and so forth.

A cloud contains water, water vapor, a great amount of water microdroplets, little ice crystals, graupels, hail that normally are not present in neutral gases nor plasmas, atmospheric air with its regular composition, anionic species, free electrons, free protons, and the charged constituents of cosmic rays that traverse it. These elements carry a lot of the charge. Yet, for the most part, a cloud is nonconducting and its dielectric components make up the great proportion (> 93%) of its composition. We consider that our cloud under consideration contains a great number of minuscule pollutant particles. Pollutants in clouds are important because they contribute to attachment of free charge particles and to its dielectric character [6, 7].

A cloud is always moving and is in vertiginous motion in all directions, and many ice crystals, anions, cations, graupels, and water droplets are in the upward, downward, and turbulent air currents. Sometimes, it contains strong air currents forming divergences, curls and, of course, turbulence, and very importantly, also, vertical upward and downward hydrodynamic currents, and they can be, frequently, kilometers tall, as is the case of cumulonimbus. These constituents move, friction, collide energetically, and a multitude of charge separation and charge transfer processes take place and have been identified, quantified, and reported for decades now [10–32]. A cloud is always immersed on the solar ultraviolet bath, the cosmic rays, natural radioactive decay earth's products and the natural electric field present in the atmosphere at all altitudes.

5. Charging mechanisms: charge separation in electrified clouds

Clouds, being mainly nonconductive, can become electrified, and they sustain frequently very large quantities of electric charge, both positive, ρ^{+}, and negative, ρ^{-}. They become electrified-charged by motion, collisions, fracture and friction of ice little crystals, water droplets, graupel snow, air molecules, and so on with tiny ice crystals (they end up with charge in their surface). Another electrification mechanism is by experiencing atmospheric electric fields, just polarizing atoms and molecules, and even breaking down some of these polarized molecules/atoms. Radioactive products can produce charged species, X-rays, and gamma rays and can dislodge electrons from their atoms/molecules. Then, we have free electrons on the one hand and cations on the other hand. Ultraviolet-visible (UV-vis) radiation from the sun and cosmic gamma rays can contribute, daily, continuously, to the production of free charge in clouds.

Very recently [61], it has been found that water microdroplets show large electric fields just outside their surface and uncompensated charge resides on their surface [62], no collisions, nor radiation needed. It is a natural phenomenon that charge concentrates, uncompensated, on the surface of these water microdroplets. It is also known that small ice particles carry charge (or equivalently, radical species) on their surfaces [10–12]. The exact production processes of this charge on these water particles continue to be investigated at the present time. Pollutant molecules also show the ability to catch, trap, on their surfaces free charge, anions and/or cations, as already mentioned.

Now, in clouds, there might appear, for short periods of time, zones showing some conductivity, σ, and this means charge mobility at least for short intervals of time appear concentrated stochastically here and there [63].

In summary, the charged elements in an electrified, nonconductive, cloud include graupel, large and small, water droplets, ice crystals, some in needle-like form, cloud drops, and ionic species normally present in the atmosphere, charged pollutant species, free electrons, free protons, and the charged constituents of cosmic rays and the charged products they produce within the domain of a cloud. And these charges are produced by several mechanisms that include, but are not limited to, noninductive processes such as collisions, breakage, fragmentation, and inductive processes such as polarization of molecules, redistribution of charge on the surface of water microdroplets, electric fields attracting and concentrating heavy charged-particles, and so on.

Hence, electrified clouds display definite regions that concentrate electric charge, positive in some regions, ρ^+, and negative in others, ρ^-. Then, most clouds posses a rather complicated charge structure, electric field lines emanate from ρ^+ regions (sources) and sink into ρ^- regions (sinks). Hence, the charge structure, the electric fields, $\vec{E}$, they produce with complex electric field line maps, and their dielectric character along with their conductivity (spots-like) property provide an electrified

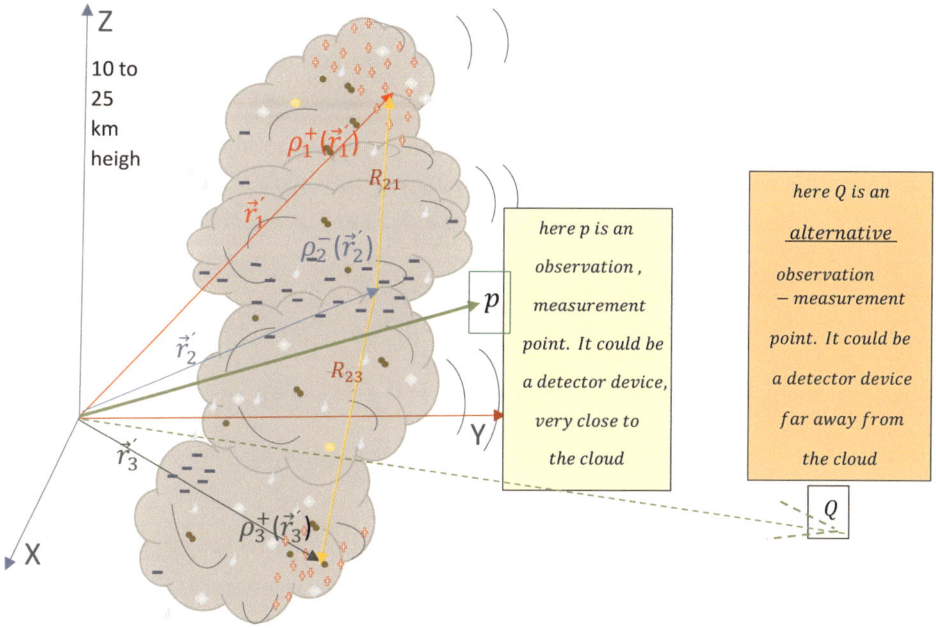

Figure 4.
Essential electromagnetic features of a moving electrified cloud include quantitative features: (A) vector positions of its charge distributions; (B) distances between the "centers" of charged regions; (C) specific regions of accumulation of charge; and (D) position of observation measurement points, p close to the cloud and Q far away from the cloud. It shows some of its basic elements such as water microdroplets, small pieces of ice, dipolar molecules, and small conducting inclusions. The electromagnetic heterogeneity of the cloud and the accumulation of charge, ρ^+, ρ^- are enhanced. The location of an observation point (or measurement location) is $\vec{r}_p$ (green). The location of ρ_1^+ is given by the vector position $\vec{r}\,'_1$, the location of ρ_2^+ is given by the vector position $\vec{r}\,'_2$, and the location of ρ_3^- is given by the vector position $\vec{r}\,'_3$, and the prime on $\vec{r}$ always indicates that we are dealing with the position of source charge, which in turn is the source of the electric field and the electric potential this cloud carries. Any conceivable charge structure in a cloud is denoted in this work by $\{\rho^+, \rho^-\}^{TOT}$, and it is an unknown of the system.

cloud with very particular electromagnetic properties that nonconducting gases do not have. Below, we go into some detail into these electromagnetic properties.

These charges can form long and wide charge distributions, some quasi-planar, others more volumetric, which locations are labeled as $\vec{r}'$, and these charges tend to accumulate in more or less well-defined regions inside the cloud; see **Figure 4** for details of a typical electrified cloud structure.

6. Distribution of free electric charge in clouds

The distribution of free electric charge in clouds is positive in one or more regions and negative in other regions. **Figure 4** shows such charge and spatial configuration and vector distances. Our cloud, any cloud that can sustain free charge in it, in any possible charge structure configuration, can have distributions of charge, let say:

$$\rho\left(\vec{r}\right) = \begin{cases} \rho_1^+\left(\vec{r}_1'\right), & \vec{r}_1' \text{ source charge Region 1} \\ \rho_2^+\left(\vec{r}_2'\right), & \vec{r}_2' \text{ source charge Region 2} \\ \rho_3^-\left(\vec{r}_3'\right), & \vec{r}_3' \text{ source charge Region 3} \end{cases} \tag{5}$$

similarly for $\rho_4^-(\vec{r}_4')$, $\rho_5^-(\vec{r}_5')$, and so on. The electric fields produced, per unit volumen of charge, are $\vec{E}_i$ = $K\rho_i\,(\vec{r})\hat{r}/r^2$, where $k = 1/4\pi\epsilon_0$ and i = 1, 2, 3, 4, 5, ... If the distribution is discrete, $\rho^{\pm}\left(\vec{r}'\right) = (1/V)\sum_i^N q_i$, or the corresponding integral if the density of charge is modeled continuosly. We locate the charges in the cloud and indicate a reference system as shown in **Figure 4**. Then, the electric field per unit volume of charge is at the observation point:

$$\vec{r}_2, \begin{cases} \vec{E}^+\left(\vec{r}_p,t\right) = \mathrm{k}\left(\rho_1^+\left(\vec{r}'_1\right)/R_{1p}^{\;2}\right)\hat{r}_{12} \text{ for } \rho_1^{\;+}\left(\vec{r}_1'\right) \\ \vec{E}^-\left(\vec{r}_p,t\right) = \mathrm{k}\left(\rho_2^{\;-}\left(\vec{r}_2'\right)/R_{2p}^{\;2}\right)\hat{r}_{2p} \text{ for } \rho^-{}_2\left(\vec{r}'\right) \\ \vec{E}^+\left(\vec{r}_p,t\right) = \mathrm{k}\left(\rho_3^+\left(\vec{r}'_3\right)/R_{3p}^{\;2}\right)\hat{r}_{3p}, \text{for } \rho_3^{\;+}\left(\vec{r}_3'\right) \end{cases} \tag{6}$$

$|\vec{R}_{ip}| = R_{ip} = \vec{r}_p - \vec{r}_i$, and $\vec{r}_1 = \vec{r}'$ is the source/sink region where $\rho_{free}^{\pm}$ is located and $\vec{r}_p$ are positions of the observer (where the field has been evaluated). The total field at the observation point "p" is the sum of the fields in (6).

Partial information on $\rho\,(\vec{r}')$ would produce only partial information on $\vec{E}$, its divergence, and its electric energy, $W \propto \left|\vec{E}\right|^2$, and an electromagnetic analysis is incomplete. The measurements of electric fields have given valuable information on the average values and their quite large variations-fluctuations. Conventional breakdown field values (3000 kv/m) are known as the breakeven field (202 kv/m) and the runaway field (280 kv/m) values [10, 12, 25].

We must note that $\rho(\vec{r}')$ in this work always represent an already electrified, or charged region in a cloud irrespective of its extent, its particular mathematical form, being it a discrete conglomerate of charge, or a “modeled” continuous distribution of charge, ρ^+ indicates a distribution, or extension, or structure of positive charge from which divergent electric field lines emanate, and ρ^- indicates a distribution of negative charge in which the electric field lines converge, sink. Any conceivable charge structure in a cloud is denoted in this work by $\{\rho^+, \rho^-\}^{TOT}$, and it is an unknown of the cloud system. It is extremely hard to determine $\rho\,(\vec{r}\,')$. Observations for decades have given only partial knowledge of it. Nowadays, a fuller, more detailed determination of $\rho\,(\vec{r}\,', t)$ and of the electric field, $\vec{E}(\vec{r}_p, t)$, it produces at the observation position $\vec{r}_p$ continue to be at the top of the electrified cloud-lightning research. Another important electromagnetic property of clouds to be considered is its dielectric character, and we go into it, next.

Clouds contain dielectrics. Some of the molecules and atoms that compose a cloud are dielectric (some very strong dielectric and some weak dielectric), some other components are already electrified, and others can be electrified by noninductive mechanisms. Let us look at these dielectric molecules (which produce bound charges) with a bit more detail. Let us think just on water molecules, water microdroplets, pollutant particles, and little aggregates of H_2O). $\rho(\vec{r}\,')$ will apply an electric field $\vec{E}\left(\vec{r}_d, t\right)$ to these molecules at their position, $\vec{r}_d$, at time t and produce a separation of the centers of positive charge from the center of negative charge, producing an atomic/molecular dipole moment given by $\vec{p} = q\vec{\ell}$. For thousands and thousands of these dielectric molecules lumped, packed, together and experiencing the same electric field $\vec{E}$, a macroscopic, total, electric moment is formed $\vec{\mathcal{P}} = \sum_i^n \vec{p}_i$, *with* $n \geq 10^4$ or much more greater, reaching nano-, micro-, millimetric, and/or larger dimensions. Then, a macroscopic density of electric moment is constructed, $\vec{P} = \vec{\mathcal{P}}/\text{vol}$ [7–9]. When $\vec{P}$ is not distributed uniformly in the volume of the aggregate cloud, which is a likely scenario, then $\vec{P}$ becomes divergent and $\text{Div}\cdot\vec{P} \neq 0$; hence, a volumetric bound charge, $\rho_b = \text{Div}\cdot\vec{P}$, appears. In addition, a surface-bound charge, $\vec{P}\cdot\hat{n}|_{\text{surface}} = \sigma_b$, appear on all the boundaries of the dielectric parts of the molecule, or aggregate, and this can happen also to water microdroplets [62]. ρ_b and σ_b can only be determined, theoretically, when the polarization vector $\vec{P}$ is known. Experimentally, the measurement of the electric field is just outside the dielectric molecule, and aggregate or microwater droplet will allow us to determine $\vec{P}$ and the volumetric and surface distributions of the bound charges. This knowledge will help us to understand their electric behavior in front of other charges [7, 8] and/or immersed in electric fields.

In real systems such as clouds, thunderclouds, plasmas, dielectrics in clouds, or for electronics, to know the distribution of free charge is challenging by itself, to know the distribution of bound charges is even more challenging, then we would need to calculate or evaluate both, the electric field vector, $\vec{E}$, and the macroscopic polarization vector $\vec{P}$. However, using the first Maxwell equation in the form, $\text{Div}\cdot\vec{E} = \rho_{free}/\epsilon$, we need to know in advance only ρ_f (and not ρ_b, this is a relief), the total electric permitivity, $\epsilon = \epsilon_o\epsilon_r = \epsilon_o(1+\chi_E)$ of the medium, then the vector field $\vec{D} = \epsilon\vec{E}$ is evaluated and from it, $\vec{E} = \vec{D}/\epsilon$. The algebraic process is simple, and it applies equally well to a molecule, an aggregate of dielectrics, a cloud region, or a complete cloud.

When the dielectric medium is sufficiently diluted and it extends for some micrometers or for kilometers, then ϵ could approach ϵ_0, and we can consider to be working with a standard clear air (in the lab, or outdoors with unpolluted air, etc.) or as when we have free charges in vacuum. Otherwise, some consideration should be paid to the polarizability of the medium since $\vec{P} = \epsilon_o \chi_E \vec{E}$. For linear and homogeneous dielectric materials, the three vectors go along the same directions, but when the dielectrics are not linear, $\vec{P}$ and $\vec{E}$ are no longer colinear and the response becomes more elaborate [7–9]. The existence of the vector $\vec{P}$ brings about an electric field produced by $\vec{p} = q\vec{\ell}$, that is, $\vec{E}p$ that opposes in direction to the original $\vec{E}$ (produced by free charges, ρ_f); hence, the total electric field in the midst of the dielectric portions of the system is reduced from $\vec{E}_i$ to, $\vec{E}_{TOT} = \vec{E}_i - \vec{E}_p$ and breakdown of the dielectric constant would be more difficult to achieve, and the transport of some free charge and free currents, $\vec{J}_f = \frac{1}{A}\frac{dq}{dt}\hat{n}$, through the dielectric regions would be more difficult to establish and of less magnitude. Hence, let us suppose we have a cloud region with free charge density ρ^+ facing another free charge density ρ^- with dielectric molecules, aggregates between them, but not totally filling the space in between; hence, there will be small regions, filaments, and islands, of nondielectric matter. The zones more clearly dielectrically occupying most of the cloud volume (let us say <90%), even covering space in between the charged regions, would leave only small regions, islands, and filamentary spaces with $\vec{p} \to 0$ approaching the free space condition in which $\vec{E}$ p $\to 0$, and the value of electric field is $\vec{E}_i$ not reduced by $\vec{E}p$. So, these nondielectric spaces would be more favorable for electric currents, $\vec{J}_f$, to form and flow. It is, already, very well known that dielectric matter is not a favorable medium to sustain electric currents [7–9]. If polarizable molecules in the cloud are moving, as they are, the possible trajectories of electric currents also change.

So, in a cloud the total charge is, ρ_{tot}, and it is composed of free charge produced by ionization, application of an electric field "$\vec{E}_i$", vertiginous vertical upward/downward hydrodynamic currents, friction, collisions, cosmic rays, solar radiation, etc., and bound charge produced by polarization, $\vec{P} = \epsilon_o \chi_E \vec{E}$, of molecules in the medium, provoked by the applied $\vec{E}_i$ field.

So, for the above, we want to consider free charge only: if we want to consider the structure, distribution, of free charge only, that is $\{\rho^+, \rho^-\}^{free} \equiv \rho_f$, this quantity is more amenable to work with Maxwell equations. For example, Gauss Law $\nabla \cdot \vec{E} = \rho_f / \epsilon$ can be put in terms of ρ_f, and then we will be dealing with sources or sinks of electric field lines, but only of free charge. And if not known, it can be modeled and/or inferred (partially) from measurements.

If, in addition, the cloud contains no magnetizable elements, then the magnetization vector, $\vec{M}$, is zero, and if the Maxwell displacement current is very small compared to the free currents in our turbulent thundercloud, then we neglect the term $\frac{\epsilon \partial \vec{E}}{\partial t}$ in (4) and we are left with the reduced Maxwell equations, (7)–(10) for this kind of electrified cloud with dielectric material in it and some free charge able to move, and considering that a cloud is a very heterogeneous structure changing very rapidly in time:

$$\mathrm{Div} \cdot \vec{E} = \rho_f / \epsilon \tag{7}$$

$$\mathrm{Rot} \times \vec{E} = -\frac{\partial \vec{B}}{\partial t} \tag{8}$$

$$\mathrm{Div} \cdot \vec{B} = 0 \tag{9}$$

$$\mathrm{Rot} \times \vec{B} = \left(\vec{J}f + \frac{\partial \vec{P}}{\partial t} \right) \tag{10}$$

The constitutive equations are $\vec{P} = (\epsilon - \epsilon_o)\vec{E}$, $\vec{D} = \epsilon\vec{E}$, and $\vec{J} = \sigma\vec{E}$, and conductivity in the regions that sustain plasma like matter could be modeled as before [33]:

$$\sigma = e^2 \left({}^{N^+t^+}/_{m^+} + {}^{N^-t^-}/_{m^-} \right) \tag{11}$$

where all quantities, for our very dynamic cloud, are very strong functions of space and time. The particular values of the parameters and charge distributions and electric fields at time t_1 in a locale A are not necessarily the same at the same locale at time $t_2 > t_1$, with $t_2 - t_1 \ll 1$ millisecond, and this leaves the theoretical treatment with Maxwell equations as "snapshots" of the resulting fields, currents, etc., as quasistatic frames as time evolves.

7. Clouds interacting with more clouds and its environment

A cloud with a typical charge structure, electric field configuration, with its dielectric and conducting character and in the presence of other charged clouds and electromagnetic environment was schematically shown in **Figure 2**. The distribution of free charge with positive charge on the upper part of the cloud ρ_1^+ is the source of divergent electric field lines. Negative, extended, distribution of charge in the middle section of the cloud, ρ_2^-, is a large, convergent sink of electric field lines, and another smaller distribution of positive charge in the lower part of the cloud and somewhat toward one side, ρ_3^+, left, or right, were located in a reference frame, and the electric fields they produced were given also in **Figure 4**. Metallic towers, an airplane flying nearby, a mountain, trees, ground, and lakes are also considered. The central cloud could be very tall and active with wind divergences and curls, upward convection currents, then downward and turbulence, and cumulonimbus clouds are just one example.

This cloud has also other neighboring clouds, having at least another electrified cloud, $\rho_{near}^{\pm}$ nearby that can be electrified (carrying even more charge and producing more electric fields that move some other free charge) and energetic exchange of their charge can occur, sometimes accompanied with lightnings. Below the continental clouds, there is always ground, which is for the most part charged positively. Above the cloud there is another charged volumetric region and with conductivity, the ionosphere, σ, just as mentioned before.

More about clouds interacting with clouds: A cloud is not only a reservoir of great amounts of water, ice, water vapor, atmospheric gases, and electrical charge-producing electric fields. It is a tremendously dynamic structure that contains megas and megas of Joules of hydrodynamic energy and gigas and gigas of Joules of electromagnetic energy. The energy, W, of a configuration of charges in an electrified cloud with n charges located at, mutual, distances $r_{ij} \approx r_{jk}$ in a quasistatic ensemble is $W = k\sum_{i<j} q_i q_j / r_{ij} = q_N V_{N-1}$ with its electric potencial $V_{N-1} = K(q_1/r_{1N} + q_2/r_{2N} + \frac{q_3}{r_{3N}} + \frac{q_4}{r_{4N}} + \ldots q_{N-1}/r_{(N-1)N})$.N could reach trillions and trillions of charged particles. Such energy is released and channeled in accord with the electromagnetic environment, EME. EME is composed of ground, mountains, trees, other electrified clouds, dielectric media, sea surface, ionosphere, temperatures, pressures, pollutants (varying with height), planes flying near or inside clouds, and metallic structures (be them tall, small, tubular, flat, and so forth). For an electromagnetic study of an electrified cloud, it is indispensable to know some physical parameters of its EME. For example, just to mention a couple of examples, the ground is a huge reservoir of positive and negative charge probably, mainly, due to the great amount of minerals, nutrients, dirt, and water in it. The ionosphere is a fully conducting medium with a lot of free charge (mainly electrons) in it. An adequate electromagnetic treatment of an electrified cloud should include its EME. The standard atmosphere has also plenty of electric charge and not only in the ionosphere that is up high several kilometers (around 90 km) above ground, and this air region is continuously exposed to cosmic radiation, the sun's ultraviolet radiation, and the electric fields of electrified clouds. Hence, molecules and atoms get energetically excited, "broken down", and ionized, captions and anions are formed, and free charged particles, mainly electrons and protons, are produced and constitute the charge in that gaseous medium.

8. Two cloud charge structure models that can produce high electric fields in their vicinity just momentarily and locally

In this section, we construct two simple cloud charge structure models that can produce high electric fields in their vicinity just momentarily and locally, schematically presented in **Figures 5** and **6**. Within these, a discharge process is being cooked (build) up. Many charge densities-structures in clouds $\rho_1^+, \rho_2^-, \rho_3^{++}, \rho_4^-, \rho_5^{--}$, etc., are possible. Moreover, clouds are, by no means static structures, they move, change form in their insides and outsides, in question of microseconds, by internal currents (up and down hot-cold streams) and because of the wind. This makes the clouds and their charge densities in very dynamic structures.

Within this dynamics, let us suppose that the charge densities move such that "pointed" distributions of charge, $\breve{\rho}^+, \breve{\rho}^-$ appear (and dissappear) in a region *a* of a cloud, then hot spots of large electric fields, $\vec{\breve{E}}$, and large divergences, DIV$\cdot \vec{\breve{E}}$, emerge for just a fraction of a second. Also, in another close by cloud or another region b in the same cloud, other charge distributions, $\overline{\overline{\rho}}\,+$, and $\overline{\overline{\rho}}\,-$, can get closer, kind of capacitor type of structure, due, again, to strong winds and rotationals and upliftings (down-liftings), creating, again, a zone with very high electric field, $\vec{\overline{\overline{E}}}$. Both scenarios are shown in **Figures 5** and **6**, respectively. Other charge structures are possible that elicit, locally and momentarily, high electric fields. The creation of strong electric fields within electrified clouds is a pre-requisite for discharging and formation of leaders, streamers, and lightning, the appearance of nonrelativistic and relativistic moving electrons, hence, the rapid increase of conductivity, emission of gamma and X-rays, and so on. We see in more detail these two possibilities now.

We envision here, case A) the emergence of electric field hot spots as a result of, by chance, the formation of pointy charge "processes", $\breve{\rho}^+$, and $\breve{\rho}^-$, coming out from some charged regions, as shown in **Figure 5**. This makes $\overrightarrow{\breve{E}}$ larger close to them and DIV$\cdot\overrightarrow{\breve{E}}$ also larger; hence, more field lines get out of the positive "pointy" charge density and close to it more field lines cross a unit area.

More field lines enter into $\breve{\rho}^-$, sink, and close to them more field lines converge, attracting more positive mobile charges. This, in turn, increases the potential difference between these regions and regions of less $\rho^{\pm}$. The electric force that a single charge q of mass m can experience in those electric field hot spot regions gets, consequently, larger, and if $\vec{F} = q\overrightarrow{\breve{E}}$, its acceleration is $\vec{a} = q\overrightarrow{\breve{E}}/m$. If we disregard collisions for the moment and consider an enlarged-effective mean free path, ℓ, for these moving charges, then there are not, by definition, important damping processes that stop this increasingly accelerated motion. We take an enlarged effective mean free path as a composed trajectory in which collisions and gazing scatterings do not, appreciably, affect greatly the acceleration, nor the kinetic energy of the particle and the continuous work of the electric field, $W = q\vec{E}\cdot d\vec{\ell}$, keeps moving the charge.

In scenario B, $\bar{\bar{\rho}}+$ gets much closer to $\bar{\bar{\rho}}-$ as pictured in **Figure 6**. A kind of capacitor is formed for which the $\bar{\bar{\rho}}+$ distribution gets much closer to $\bar{\bar{\rho}}-$ due to wind

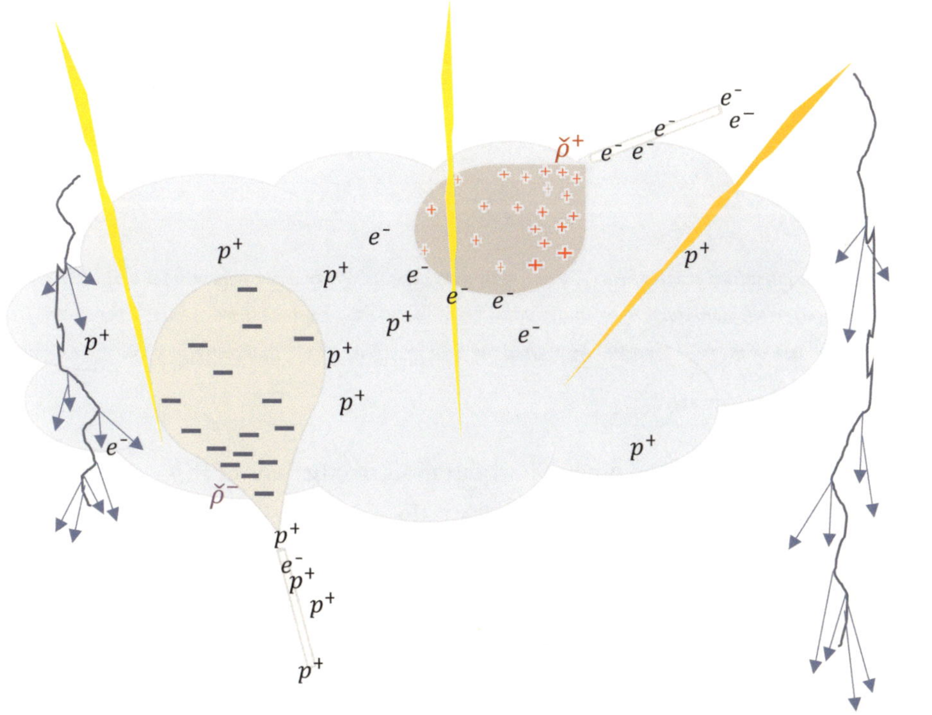

Figure 5.
Pointy charge structures, $\breve{\rho}^+$, and $\breve{\rho}^-$ developed in a moving cloud. Some important redistribution of free and bound charge starts to take place. Free charge, e^-, p^+, anions, and/or cations already in the body of the cloud will get strongly attracted to these force field lines and will move accordingly, as $\vec{F} = q\vec{E} = m\vec{a}$, and then they will accelerate as, $\vec{a} = q\vec{E}/m$. they continue to accelerate as long as collisions do not stop them. This regime is called enlarged-effective mean free path, ℓ.

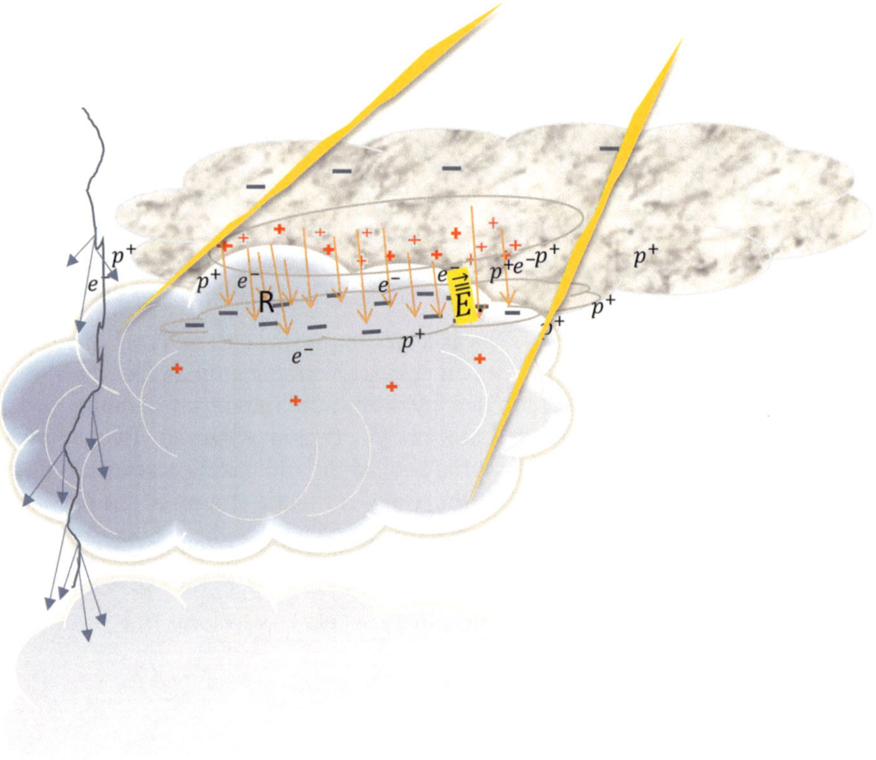

Figure 6.

A momentarily intensified electric field. A scenario in which a kind of capacitor is formed for which the $\bar{\bar{\rho}}^{+}$ *distribution gets much more closer to* $\bar{\bar{\rho}}^{-}$ *due to wind forces.* $\vec{E}$ *increases between them as* $[1/r]^2$*, the electric field in between is the sum of the field produced by* $\bar{\bar{\rho}}^{+}$ *and the field produced by* $\bar{\bar{\rho}}^{-}$*, and charged species are attracted to this* $\vec{\bar{\bar{E}}}$ *field and then accelerated.*

forces that push one to the other. Short duration fluctuations of $R(\rho^{+}, \rho^{-})$ in **Figure 6** increases $\vec{E}$ between them as $[1/R(\rho^{+}, \rho^{-})]^2$, the electric field in between is the sum of the field produced by $\bar{\bar{\rho}}+$ and the field produced by $\bar{\bar{\rho}}-$, it becomes very high and then again, it attracts free charge that happens to be in that space, or close by at that moment, t*. The Coulomb force they feel is, again, $\vec{F} = q\vec{\bar{\bar{E}}}$, and its acceleration is $\vec{a} = q\vec{\bar{\bar{E}}}/m$. They accelerate with no damping inside the elongated effective mean free path, ℓ. This increasingly accelerated motion can continue without being slowed down in this regime. We propose that the electric work made by the $\vec{\bar{\bar{E}}}$ field, $W = q\vec{\bar{\bar{E}}} \cdot d\vec{\ell}$ is large enough to keep some, but not all, electrons accelerated, these electrons will become runaway electrons in the next stage, few microseconds ahead.

But, it just so happens that this motion can also be maintained and even increased by positive feedback in which the loss, by collisions, of moving energy is compensated by gain of energy in other elastic collisions in which more massive particles transfer linear momentum and energy to these lighter charged particles. Both scenarios can contribute greatly the cosmic rays, and apparently, more specifically their secondary products at lower altitudes, well below the lower layer of the ionosphere, well into the stratosphere.

If this intense electric field window, in both scenarios, last for some microseconds, a lot of charged particles can be moved on along these force field lines and can reach distances of a fraction of a kilometer or more. Then, we can talk about electric current flowing down to lower electric potential or of positive charge traveling, current, J+, toward the negative pole or "electrode," and/or of electrons traveling toward the positive pole or "electrode," conforming a negative current, J-. Of course, both kinds of currents can happen simultaneously, or sequentially. The end point could simply be the air, or the ground, or a metallic structure, a tree, even the ionosphere. We see that the liberation of electric energy W is quite large, since part of the charge structure is being disassembled, let's say, just as an order of magnitude illustration, 10^{12} charges (electron equivalents) are being "send" (ideally) to infinity (toward electric potential zero), then W will be 10^{12} times the work done to remove one single charge from the cloud charge structure to infinity [7–9]. Let us suppose this unit work is about 0.23 nano Joule, then in order to dismantle 10^{12}particles from the ensemble, the energy spend would be $(0.23\times10^{-9})\times10^{12}$Joule = $2.3\text{x}10^3$ Joule = 2.3 thousand Joules, that is quite a bit of energy. Where all this energy goes to?

The intense field values reached, in both scenarios, just before the beginning of these J+ and J- currents, and the expenditure of the above-mentioned electric energy W are defined as the "at the verge of discharge electric fields," and we label them as $\vec{\mathbf{E}}_{\varsigma}$. Then, we are here with the maximum source and sink of electric field lines in both cases. We want to concentrate more on scenario B. Notice that here $\vec{E}_{\varsigma} = \vec{E}^{+} + \vec{E}^{-}$ in the in-between zone. Since dielectrics reduce the total electric field in spaces where $\vec{P} \neq 0$, as: $\vec{E}_{TOT} = \vec{\mathbf{E}}_{\varsigma} - \vec{E}p$, then within the dielectric matter $\vec{E}$ is not increasing toward $\vec{E}_{\varsigma}$ in there. However, if the dielectric matter in the cloud is inhomogeneously distributed and $\vec{P}$ varies greatly in the interior of the cloud, v. gr., covering the space with big and small dielectric patches, then some regions in between patches are almost $\vec{P}$-free slender-tortuous elongated regions in which $\epsilon \to \epsilon_0$ and the electric field, still, is $\vec{E}_{\varsigma} = \vec{E}^{+} + \vec{E}^{-}$ with no $\vec{E}_p$ reduction. These two scenarios pinpoint very specific local-cloud "spaces" that fulfill particular conditions in which $\vec{E}$ can reach the "on the verge of discharge field," $\vec{E}_{\varsigma}$. Two more fundamental ingredients for discharge to occur are as follows: (A) the presence of a lot of free light charged particles, mainly electrons, protons, and small ions, to be transported, and (B) an increased conductivity in these locations at time t*, both go hand on hand.

Once we are right at the verge of discharge, in these locations, with an increased conductivity, and a corresponding resistivity able to produce a lot of heat from Joule effect, what happens next?? well, if in the following few microseconds charge $\rho_b^{+}\left(\vec{r}_b^{\,\prime}\right)$, and $\rho_d^-\left(\vec{r}_d'\right)$ drops and/or gets redistributed again, or gets separated, larger R values, and then $\vec{E}$ diminishes quickly and potential discharge is inhibited, as when a huge sneeze gets frustrated! and the cloud continues in the charged state. Otherwise, the discharge process, really, takes off. The above makes us think of the many inhibited discharges that are hold-on in the last microseconds, because one of the above condition ceased to exist.

9. Then the discharge "Proceeds"

Discharge occurs triggered by not-so-well-understood mechanisms, but some likely mechanisms are as follows: (A) simply large amplitude variations of electric fields; (B) cosmic ray energetic events triggering the production of bare, accelerated electrons; (C) energetic ionizations by atomic/molecular collisions; (D) breakdown of the dielectric constant of cloud-air at some locations nearby, or between ρ^+ and ρ^- cloud charged regions; and (E) increase of $\vec{E}$ intensity due to a sudden increase, surge, of charge in ρ^+ and ρ^-regions, since $\left|\vec{E}\right| \propto \rho^{\pm}$. But, it is also posible to increase $\vec{E}$ by diminishing the distances, $|\vec{r}| = r = \vec{r}_2 - \vec{r}_1$, among the charged regions. We go into more detail here. (F) E increase due to charge distance decrease. Movement of ρ^+ (ρ^-) toward ρ^- (ρ^{+-}) as in scenario B above due to a strong air dynamics fluctuation, turbulence, rotations, convection, so that, short duration fluctuations of $R(\rho^+, \rho^-)$ increases $\vec{E}$ in-between as $1/[R(\rho^+, \rho^-)]^2$, so, if R goes from R_0 to $(1/4)R_0$ then E goes from E_0 to $16E_0$ and if E_0 = 120 kvolts/m, then the increased field would be now E = 1920 kvolts/m, and this momentary field is quite enough to produce the breaking of some polarizable molecules. Some ions, cations, and free-unbound electrons are produced and will immediately get accelerated toward/away from the closest charged region $\rho^{\pm}$ of opposite sign. Even in the case of just reducing the distance by half, if R goes from R_0 to $(1/2)R_0$, then E goes from E_0 to $4E_0$ and if E_0 = 120 kvolts/m, then the increased field would be now E = 480 kvolts/m. This field is larger than the 320 kvolts/m of the reported breakdown field [25]. One of the objectives of this work has been to construct plausible scenarios, electromagnetic conditions, on which large intensity electric fields $\vec{E}$, with durations of at least some micro-, or milliseconds, can develop in the in-between and/or around the charged cloud regions, be them on the upper/center/lower part of the cloud, or toward the sideways, as shown in **Figures 5** and **6**.

In **Figure 7A**, we have imagined five charged particles at five different locations in this electrified environment (well within the cloud), a free electron, a positron, a positive ion, a proton, and a polarized molecule, experiencing the presence of the electric field $\vec{E}$, and the vector distances should properly be measured from the coordinate system arbitrarily situated at $\mathcal{O}$. All of these charges experience Coulomb force at their location. It should be pointed out that not always it is trivial (or easy) to calculate such a field $\vec{E}$. In fact, it continues to be an open problem in thundercloud systems.

Now, we will do a "Thought" experiment: If the relative distance of the field sources and the charged particles is changed, several of the forces experienced vary

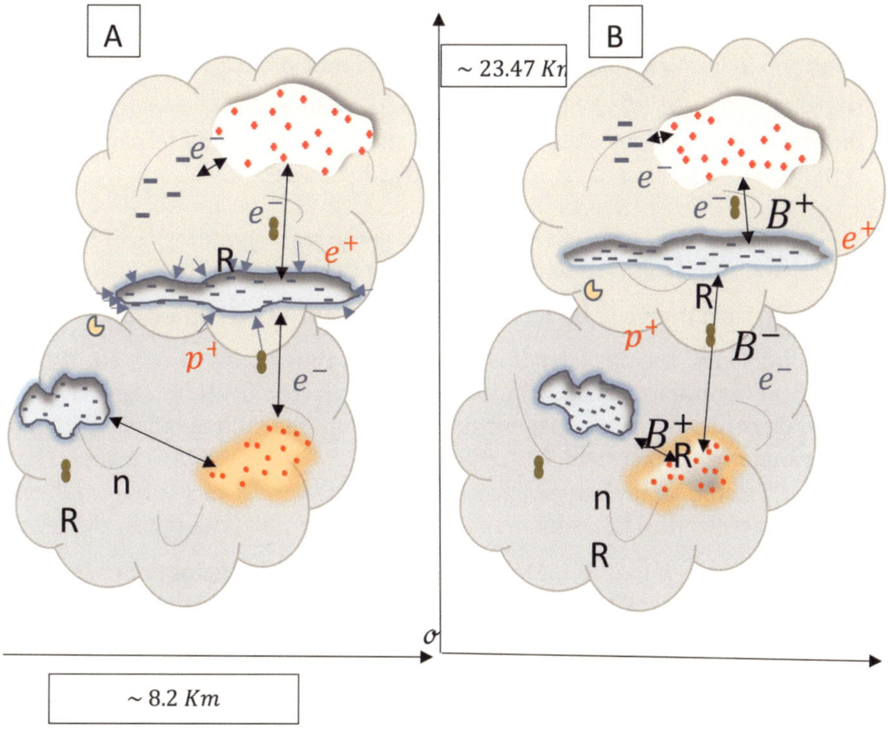

Figure 7.
We have imagined five electrical objects located at seven different positions in this electrified environment. They all experience the presence of the electric field $\vec{E}$; hence, a coulomb force depends on their location P1, P2, ..., P5, *well inside the cloud: (A) original charge configuration with typical distributions of charge and distances; (B) the charge distributions have been moved to get some closer and others farther away, the electric field and coulomb forces change accordingly. A zone of higher $\vec{E}^{+}$ field is denoted B^{+} and another region with weakened $\vec{E}^{-}$ field is denoted B^{-}.*

even drastically, the force field lines get distorted (they follow different paths now), and the electric energies will change correspondingly. The electric potential lines are also modified, but it is always obeying $\vec{E}\cdot d\vec{\ell}_{pot} = 0$, as shown in **Figure 7B**. Above shows distances modified, and electric field and electric potentials have changed with respect to panel A. Divergences and rotationals in panel A have changed, and now stronger divergences appear and other rotationals appear in panel B. For example, see the regions marked B^{+} and B^{-} with corresponding electric fields $\vec{E}^{+}$ and $\vec{E}^{-}$ and electric potentials $V^{+}$and V^{-}. These spatial changes bring about several electromagnetic consequences, some are as follows:

Region B^{+}/B^{-} with more/less force field lines concentrate on more/less electric energy (measured in Jules) per unit volume since W/vol = $(\epsilon/2)\ E^2$ and since $E_{B+} >> E_{B-}$ then $W_{B+} >> W_{B-}$. In fact, since E α $(1/r^2)$, then W α $(1/r^4)$, and the electric energy is a very strong function of the distance between charged bodies. If distance R changes just by half, R $\rightarrow$ R/2, then energy density will increase 16 times in region B^{+}.

A. Consider two identical charges $\tilde{q}$ and $\overline{q}$ (two electrons, two positrons, two protons, two anions, and two cations) of equal mass, $\tilde{m}$. We put $\tilde{q}$ in region B^{+}at a point p at half the distance $\tilde{R}$ from the sources, and the identical charge $\overline{q}$ is placed in region B^-, at a similar distance from its sources. $\tilde{q}$ experiences a much stronger Coulomb force, $\vec{F}^+ = \tilde{q}\vec{E}^+$ than the identical charge $\bar{q}$ placed in region B^-, where $\vec{F}^- = \bar{q}\vec{E}^-$. The histories of subsequent motion of $\tilde{q}$ and of $\bar{q}$ will be very different. $\bar{q}$ could eventually recombine, and $\tilde{q}$ could become relativistic and could produce an avalanche of more charged particles.

B. If allowed to move, both charges will follow the corresponding field line directions, and if allowed to move for a time interval Δt without collisions or damping, then they will be accelerated by $\vec{F}^+ = \tilde{m}\vec{a}^+ = \tilde{q}\vec{E}^+$ in region B^+, and by $\vec{F}^- = \tilde{m}\vec{a}^-$ in B^-region. Then if $E^+ = \eta E^-$ with, for example, $\eta = 2, 10, 100, 1000$ and so on (we are modeling, arbitrarily, intense electric fields, E^+, here twice, five times, one order of magnitude, two orders of magnitude, three orders of magnitude larger than the weaker field, E^-), then $ma^+ = \tilde{q}E^+ = \tilde{q}\eta E^- = \eta(\tilde{q}E^-) = \eta \mathrm{m}a^-$; hence $a^+ = \eta a^-$, so the same charged particle placed in region B^+ in a stronger electric field will accelerate η times more than if put in region B^- with the weaker electric field E^-, $E^+/E^- = \eta = a^+/a^-$. Moreover, $|\vec{F}^+|/|\vec{F}^-| = \eta$. This helps to understand the existence of slow-moving charges and fast-moving charges in the same electrified cloud. This applies equally well to light and heavy charged particles.

C. When known the intensity of the electric fields $E^\pm$, then $a^\pm = (\tilde{q}/\tilde{m})E^\pm$ can also be determined if the mass of the particle is known.

D. If we consider the enlarged-effective mean free path, ℓ, (as defined above) with no loss of energy, no friction, as if flying in free space, and if ℓ is about one meter, tens, or hundreds of meters, then charges will reach classical velocities of $v^2 = v_0^2 + 2a\ell$, or $v = v_0 + at$. As a manner of numerical example, let us suppose an electron starts this accelerated motion inside an electric field E of just 1 kvolts/m with an initial velocity, $v_0 = 1000$ m/sec or less, and $\ell = (0.1\ \text{Km})\text{L}$, where L can be less or greater than 1 is a convenient adjusting distance parameter of the mean free path, then the acceleration $a = \text{qE/m}$ is going to be $(1.6 \times 10^{-19}$ Coulomb) (1 kvolts/m)/$(9.11 \times 10^{-31}$Kg), then $a = 1.756 \times 10^{14}[\text{m}/sec^2]$.

Then $v^2 = (1000\ \text{m/sec})^2 + 2\text{x}1.756 \times 10^{14}[\text{m}/sec^2]$. (100 L) [m], taking square root we obtain, $v \approx 1.87\sqrt{L} \times 10^8$ m/sec, and this velocity is a fraction of the velocity of light, $(1.87\sqrt{L} \times 10^8)/(3\text{x}\ 10^8\ \text{m/sec}) = 0.624\sqrt{L}$, impresive! It should be noted that in this calculation the adjusting distance parameter L is just equal to 1, and the initial velocity with which the electron enters the region of $\vec{E}$ is a very small contribution and can be safely neglected. This classical calculation of acceleration and velocity is not correct for relativistic velocities but helps to give an idea of the tremendous accelerations and velocities that electrons can reach under these circumstances.

E. The above means also that a great number of electrons ($\geq 10^{10}$) with small, medium, and high velocities just drifting (going by in any direction) in the vicinity of $\vec{E}$ is going to be accelerated to reach velocities at around 0.624 c, and many slow-moving electrons just drifting by easily fullfil this initial condition.

Since $v = v_i + at$, then the time to reach the final velocity (in this elongated effective mean free path length) is t = $(v - v_i)/a = 1.87\sqrt{L}\times 10^8 - 1000)$ (m/sec)/(1.756 × 10^{14}) [m/sec2]$\cong 1.065\sqrt{L}\mu sec$, a very short time!. We can envision many electrons drifting by in the region of $\vec{E}$ of just 1000 volts/m and get this acceleration and reach a velocity 0.6234c in just a little more than a microsecond. Here, L continues to be equal to 1.

F. The gain in kinetic energy in this accelerated motion along ℓ is $K–K_0 = ½mV^2 - ½mV_0{}^2 \cong ½mV^2$ since $V_0{}^2 << V^2$, then Δk $\cong$ ½ m V^2 = 16Lx10^{-15}Joule = 99.84 L Mev. Let us suppose the mean free path is not as large as 0.5 Km, because of small energy collisions, scatterings, frictions and so on, and let us model a more realistic mean free path of a fifth of the more idealized one, then $L = (1/5)$; hence, the velocity reduction would be $V \approx 1.87\sqrt{1/5} \times 10^8$ m/sec = 0.28c, and its kinetic energy go down from 99.84 Mev to 19.97 Mev. So in a more frictional environment which reduces ℓ from 100 meters to just 20 meters, the velocity of the electron goes down to a weak relativistic regime with about 20 Mev of kinetic energy. If collisions and energy losses become even more important and the L value becomes L = 1/20, so ℓ = 100 m/20 = 5 m, then we have $V \approx = 1.87\sqrt{L} \times 10^8$ m/sec = 0.0935 10^8 m/sec and the corresponding kinetic energy is now 99.84Mev/20 = 4.99 Mev. Velocity is now not relativistic, yet it carries an important energy with it.

G. Again, the initial velocity can be, and has been, safely neglected. The acceleration would be exactly the same, and the time to travel ℓ would be the same microseconds since it does not depend on ℓ either. So, the quotient of kinetic energy gains for the case with L = 1 with respect to L = 1/5, or 1/20 is $½m\ V^2 / ½m\ V^2$L = 1/L = 5, or 20, and the shorter mean free path ℓ modeling collisions, friction, etc., impacts directly in the velocity not reaching relativistic regimes and reaching lower kinetic energies by a factor (1/L).

H. The heavier the charged particle, the less accelerated it becomes since $a\alpha\, {}^1\!/_m$, and the larger the charge, the bigger the acceleration. Electrons accelerate and reach even relativistic velocities in time intervals of a few microseconds in fields of about 1000 V/m in diluted (almost free) space, no collisions, no frictions, and no damping. Very importantly, if the particles possess already a significant initial velocity, let's say V_0 = 0.1c, then again $V_f^2 >> V_i^2$ and our calculation still holds.

I. For nonrelativistic motions let us take $V_f = 10^{-2}(3 \times 10^8 \frac{m}{sec})$ = 0.01c, and V_i = 0.001c, then $[V_f{}^2 - V_i{}^2]^{1/2} = [(9\text{x } 10^{10})]^{1/2}(99)^{1/2}\frac{m}{sec}$. Hence $a = (V_f^2 - V_i^2)/2S \approx (3/L) \times 10^3 \frac{km}{sec^2}$, again the acceleration is mainly determined by the final velocity reached and not by its initial velocity, and this permits all kinds of wandering and drifting particles entering the zone of acceleration to reach at the end of this interval similar accelerations in similar intervals of time. This works as a velocity-acceleration funnel that also directs the "charged beam" into a particular, well-defined direction, the direction of $\vec{E}$. If the beam does not encounter many "obstacles" (collisions and scatterings), then we have an electrical current traveling some tens or hundreds of meters down that trajectory.

J. Also, $d\vec{P}^{+}/dt = \vec{F}^{+} = q\vec{E}^{+}$, and $d\vec{P}^{-}/dt = \vec{F}^{-} = q\vec{E}^{-}$, and $(dP^{+}/dt)/(dP^{-}/dt) = E^{+}/E^{-} = \eta =$ 2, 10, 100, 1000. Hence, the charge q moving in region B^{+} increases its linear momentum per unit time much more than the charge moving in region B^{-}. Under collisions and scatterings, the charges moving in B^{+} region can transfer much more linear momentum and energy to other charges and "spread" motion and energy to other charged particles, moving classically or relativistically.

K. All these points from (A) to (K) indicate in one way or another the powerful effect that a larger electric field and its direction has on N charges moving in it as compared with the same charges moving in weaker electric fields.

L. From (A) to (K), seems to us, that are conditions for a lightning starting in the cloud and directed to any number of possible endings and directions.

More dynamics and electromagnetic features can come out from the above basic calculations that can be analyzed and contrasted to field measurements.

10. Conclusions

Different transient charge distributions have been considered and their electric fields discussed, and these quantities are, for the most part unknowns, still we extracted valuable information from their vectorial properties, working within close distances, on the structures that produce pronounced sinks/sources with large divergences of $\vec{E}$ and focusing on observation points close to them. We introduced the concept of "at the verge of discharge electric field, $\vec{E}_{\xi}$ produced by two specific charge configurations". The role the dielectric character of the cloud plays in allowing/not allowing discharge was analyzed. We profiled the conditions to be met by $\vec{E}_{\xi}$ to appear and to be sustained for at least a few milliseconds. A simple, typical, electrified cloud model was constructed and described in detail. Then a few basic calculations are carried out for a thought experiment, moving charges around inside this model cloud, and we calculated velocities, accelerations, and kinetic energies for long- and short-elongated effective mean free paths, fast and slow electrons resulted in this dynamics. The acceleration of charges works irrespective of initial velocities, what is important is a light mass and an intense electric field. We analyzed what we learn from it.

Author details

Rafael Zamorano Ulloa
Superior School of Physics and Mathematics, IPN, Mexico City, Mexico

*Address all correspondence to: davozam@yahoo.com

References

[1] Ball Lightning Wikipedia [Internet]. 2022. Available from: https://en.wikipedia.org/wiki/Ball_lightning [Accessed: 18 December 2022]

[2] Saint Elmo's Fire Wikipedia [Internet]. 2022. Available from: https://en.wikipedia.org/wiki/St._Elmo%27s_fire [Accessed: 18 December 2022]

[3] List of thunder gods Wikipedia [Internet]. 2022. Available from: https://en.wikipedia.org/wiki/List_of_thunder gods [Accessed: 18 December 2022]

[4] Zeus Wikipedia [Internet]. 2022. Available from: https://www.britannica.com/topic/Zeus [Accessed: 18 December 2022]

[5] Peeters SA, Mirpour S, Kohn C, Nijdam S. A model for positive Corona inception from charged ellipsoidal thundercloud hydrometeors. JGR Atmospheres. 2022;**127**:e2021JD035505. DOI: 10.1029/2021JD035505

[6] Qixu M, Andrew G, Winn W, et al. Hidrometer charges observed below an electrified cloud. Journal of Geophysical Research. 2007;**112**:D13207. DOI: 10.1029/2006JD007809

[7] Lorain P, Corson D. Electromagnetic Fields and Waves. 2nd ed. New York: WH Freeman and Company; 1970. p. 696. ISBN: 0-7167-0331-9

[8] Feynman RP, Leighton RB, Sands M. The feynman lectures on physics. Mainly electromagnetism and matter. In: The New Millennium Edition. Vol. II. New York: Basic Books; 2010. p. 566. ISBN: 978-0-465-07998-8

[9] Jackson JD. Classical Electrodynamics. New York, London: John Wiley & Sons, Inc.; 1962. p. 656

[10] Uman MA. The Lightning Discharge 1987. 1st ed. United Kingdom, London: Elsevier, Academic Press; eBook ISBN: 9780080959818

[11] Betz HD, Schumann U, Laroche P. Book lightning: Principles, instruments and applications. In: Review of Modern Lightning Research. 1st. ed. Germany: Springer Dordrecht; 2009. DOI: 10.1007/978-1-4020-9079-0

[12] Leblanc F, Aplin KL, Yair Y, et al., editors. Planetary Atmospheric Electricity. 1st Ed. NY, USA: Springer; 2008. ISBN-978-0-387-87663-4

[13] Hare BM, Uman MA, Dwyer MAJR, et al. Ground-level observation of a terrestrial gamma ray flash initiated by a triggered lightning. Journal of Geophysical Research. 2016;**121**:6511. DOI: 10.1002/2015JD024426

[14] Dwyer JR, Schaal MM, Cramer E, Arabshahi S, Liu N, Rassoul HK, et al. Observation of a gamma-ray flash at ground level in association with a cloud-to-ground lightning return stroke. Journal of Geophysical Research. 2012; **117**:A10303. DOI: 10.1029/2012JA017810

[15] Arnone E, Bór J, Chanrion O. Climatology of transient luminous events and lightning observed above Europe and the Mediterranean Sea. Surveys in Geophysics. 2020;**41**:167. DOI: 10.1007/s10712-019-09573-5

[16] Torsten N, Ostgaard N, Reglero V, et al. A terrestrial gamma-ray flash and ionospheric ultraviolet emissions powered by lightning. Science. 2020;**367**: 183. DOI: 10.1126/science.aax3872

[17] Anjing H, Gaopeng L, Hongbo Z, et al. Locating parent lightning strokes of sprites observed over a mesoscale

convective system in Shandong Province, China. Advances in Atmospheric Sciences. 2018;**35**:1396. DOI: 10.1007/s00376-018-73064

[18] Liu F, Zhu B, Lu G, et al. Observations of blue discharges associated with negative narrow bipolar events in active deep convection. Geophysical Research Letters. 2018;**45**: 2842. DOI: 10.1002/2017GL076207

[19] Fukunishi H. Elves lightning induced transient luminous events. Geophysical Research Letters. 1996;**23**:2157. DOI: 10.1029/96gl01979

[20] Kelley NA, Smith DM, Dwyer JR, Splitt M, Lazarus S, Martinez-McKinney F, et al. Relativistic electron avalanches as a thunderstorm discharge competing with lightning. Nature Communications. 2015;**6**:7845. DOI: 10.1038/ncomms8845

[21] Chilangarian A, Hovsepyan G, Elbekian A, et al. Origin of enhanced gamma radiation in thunderclouds. Physical Review Research. 2019;**1**: 033167. DOI: 10.1103/ PhysRevResearch.1.033167

[22] Tsalkou YA, Tikhomirov VV, Marozava HP. Relativistic runaway electron breakdown and terrestrial gamma ray flashes GEANT4 simulation. 2013. arXiv:1301.5506v2

[23] Yuasa T, Wada Y, Enoto T, Furuta Y, et al. Thundercloud project: Exploring high-energy phenomena in thundercloud and lightning. Progress of Theoretical and Experimental Physics. 2020;**2020**:103H01. DOI: 10.1093/ptep/ptaa115

[24] Brunetti M, Cecchini S, Galli M. Gamma-ray bursts of atmospheric origin in the MeV energy range. Geophysical Research Letters. 2000;**27**:1599. DOI: 10.1029/2000GL003750

[25] Dwyer JR, Uman MA. The physics of lightning. Physics Reports. 2014;**534**: 147-241. DOI: 10.1016/j.physrep.2013. 09.004

[26] Zhang H, Lu G, Lyu F, et al. First measurements of low frequency sferics associated with terrestrial gamma ray flashes produced. Geophysical Research Letters. 2020;**47**:e2020GL089005. DOI: 10.1029/2020GL089005

[27] Pu Y, Steven A, Liu F, et al. Low frequency radio pulses produced by terrestrial gamma-ray flashes. Geophysical Research Letters. 2019;**46**: 6990. DOI: 10.1029/2019GL082743

[28] Ludin DI, Rakov VA, Syssoev AA, Bulatov AA, Hayakawa M. Decimeter-scale elevatedionic conductivity regions. Scientific Reports. 2021;**11**:18016. DOI: 10.1038/s41598-021-97321-4

[29] Gurevich AV, Medvedev YV, Zybin KP. Thermal electrons and electric current generated by runaway breakdown effect. Physics Letters A. 2004;**321**:179. DOI: 10.1016/j. physleta.2003.10.062

[30] Avila E, Martinez L, Pereyra R, et al. Measurements of size and electrical charges carried by precipitation particles during relampago field campaign. Earth and Space Science. 2022;**9**:e2022EA002. DOI: 10.1029/2022EA002407

[31] Luque MY, Nollas F, Pereyra RG, et al. Charge separation in collisions between ice crystals and a spherical simulated Graupel of centimeter size. JGR Atmospheres. 2020;**125**:e2019. DOI: 10.1029/2019JD030941

[32] Xu W, Celestin S, Pasco VP. Optical emissions associated with energetic electrons produced by stepping leaders in cloud to ground lightning discharges. Geophysical Research Letters. 2015;**42**: 5610. DOI: 10.1002/2015GL064419

[33] Griffiths DJ. Introduction to Electrodynamics. 3rd ed. Vol. 1999. New Jersey, USA: Prentice Hall; 2013. p. 576

[34] Halliday D, Resnick R, Walker J. Fundamentals of Physics. 10th ed. USA: Wiley; 2014. ISBN-10: 1118230647

[35] Purcell D. Electricity and Magnetism. 2nd ed. UK: Cambridge Univ. Press; 2011. ISBN 978-1-107-013660-5

[36] Roller D, Roller DH, Morrison P. The development of the concept of electric charge: Electricity from the Greeks to coulomb. Physics Today. 1955;**8**:15. DOI: 10.1063/1.3062120

[37] W. Thornhill. The IEEE, Plasma Cosmology and Extreme BallLightning holoscience.com|The ELECTRICUNIVERSE®2006. Available from: https://www.holoscience.com/wp [Accessed: 18 December 2022]

[38] Thunderstorm Effects on the Atmosphere-Ionosphere System (TEA-IS). European Science Foundation www.esf.org. 2012. [Internet] Available from: http://archives.esf.org/fileadmin/Public_documents/Publications/tea-is.pdf [Accessed: 18 December 2022]

[39] Lahanas M. Ancient Greece: Electricity, Magnetism and Pyroelectricity [Internet]. 2022. Available from: www.hellenicaworld.com/Greece/Science/en/ElectroMagnet.html [Accessed: 18 December 2022]

[40] Xolotl, a god of fire and lightning. Wikipedia [Internet]. Available from: https://en.wikipedia.org/wiki/Xolotl#:~:text=In%20Aztec%20mythology%2C%20Xolotl%20(Nahuatl,soul%2Dguide%20for%20the%20dead [Accessed: 18 December 2022]

[41] Zhang L, Gu Z, Yu C, Zhang Y, Cheng Y. Surface charges on aerosol particles—Accelerating particle growth rate and atmospheric pollution. Indoor and Built Environment. 2016;**25**(3): 437-440. DOI: 10.1177/1420326X16643799 ibe.sagepub.com

[42] Brock JR, Marlow WH. Charged aerosol particles and air pollution. Environmental Letters. 1975;**10**(1):53-67. DOI: 10.1080/00139307509435808

[43] Leyden jar, Britannica [Internet]. 2022. Available from: https://www.britannica.com/technology/Leyden-jar [Accessed: 18 December 2022]

[44] Serra AM, Torres PI, Avila RI. How electricity was discovered and how it is related to cardiology. Archivos de Cardiología de México. 2012;**82**:252. DOI: 10.1016/j.acmx.2012.06.003

[45] Mikerov AG. St. Petesburg St. Electrotechnical U, from history of electrical engineering: Electricity and magnetism in old. Middle Ages and Renaissance. 2012;**3**:65-67. Available from: https://www.researchgate.net/publication/281346112From_history_of_Electrical_Engineering_Electricity_and_Magnetism_in_Old_Middle_Ages_and_Renaissance

[46] Salama AIA. Encyclopedia of Separation Science, Mechanical Techniques: Particle Size Separation. 2000. Available from: https://www.sciencedirect.com/topics/chemistry/electrostatic-charge#:~:text=1%20Electrostatic%20charges,through%20a%20copier%20or%20printer [Accessed: 18 December 2022]

[47] Historical Introduction to learn electricity [Internet]. Available from: https://farside.ph.utexas.edu/teaching/316/lectures/node12.html [Accessed: 18 December 2022]

[48] Early History of Electricity and Magnetism. NASA [Internet]. Available

from: https://pwg.gsfc.nasa.gov/Electric/-E14-history.htm [Accessed: 18 December 2022]

[49] The Electroscope Wikipedia [Internet]. Available from: https://en.wikipedia.org/wiki/Electroscope#:~:text=The%20electroscope%20is%20an%20early,is%20proportional%20to%20its%20voltage [Accessed: 18 December 2022]

[50] Van de Graaff generator Wikipedia [Internet]. 2022. Available from: https://en.wikipedia.org/wiki/Van_de_Graaff_generator [Accessed: 18 December 2022]

[51] Volta Alessandro Britannica [Internet]. Available from: https://www.britannica.com/biography/Alessandro-Volta [Accessed: 18 December 2022]

[52] Michael Faraday Wikipedia [Internet]. 2022. Available from: https://en.wikipedia.org/wiki/Michael_Faraday [Accessed: 18 December 2022]

[53] Beiser A. Concepts of Modern Physics. 6th ed. New York, USA: McGraw-Hill; 2003. ISBN 0-07-244848-2

[54] Liu N, Pasko VP. Effects of photoioionization on propagation and branching of positive and negative streamers in sprites. Journal of Geophysical Research. 2004;**109**: A04301. DOI: 10.1029/2003JA010064

[55] Electrolysis Wikipedia [Internet]. 2022. Available from: https://en.wikipedia.org/wiki/Electrolysis [Accessed: 18 December 2022]

[56] Sundén B, Hydrogen, Batteries and Fuel Cells, 2019 [Internet]. 2022. Available from: https://www.sciencedirect.com/topics/engineering/faradays-law-ofelectrolysis#:~text=The%20first%20law%20states%20that,electricity%20in%20different%20substances%20are [Accessed: 18 December 2022]

[57] N St. J Braithwaite. Introduction to gas discharges. Plasma Sources Science and Technology. 2000;**9**:517. DOI: 10.1088/0963-0252/9/4/307

[58] Lister GG, Waymouth JF. Encyclopedia of Physical Science and Technology. Third ed. 2003. II.B.1 Classification of Gas Discharges for Lighting [Internet]. 557-595. Available from: https://www.sciencedirect.com/topics/earth-and-planetary-sciences/gas-discharges#:~:text=Gas%20discharge%20light%20sources%20are,pairs%2C%20forming%20a%20conductive%20plasma [Accessed: 18 December 2022]

[59] Wada Y, Bowers G, Enoto T, Kamogawa M, Nakamura Y, Morimoto T, et al. Termination of electron acceleration in thundercloud by intra/inter-cloud discharge. Geophysical Research Letters Online edition. 2018. DOI: 10.1029/2018GL077784

[60] Carlson BE, Lehtinen NG, Inan US. Runaway relativistic electron avalanche seeding in the Earth's atmosphere. Journal of Geophysical Research: Space Physics. 2008;**113**:A10307. DOI: 10.1029/2008JA013210, 2008

[61] Hao H, Leven I, T. Head-Gordon can electric fields drive chemistry for an aqueous microdroplet? Nature Communications. 2022;**13**:280. DOI: 10.1038/s41467-021-27941-x

[62] Shavlova AV, Dzhumandzhia VA, Yakovenko AA. Charge of water droplets during evaporation and condensation. Journal of Aerosol Science. 2018;**123**:17-26. DOI: 10.1016/j.jaerosci.2018.05.016

[63] Ludin DI, Rakov VA, Syssoev AA, Bulatov AA, Hayakawa M. Formation of decimeter-scale, long-lived elevated ionic conductivity regions in thunderclouds. NPJ|Climate and Atmospheric Science. 2019;**2**:46. DOI: 10.1038/s41612-019-0102-8

Chapter 4

Electromagnetic Relations between Materials and Fields for Microwave Chemistry

Jun-ichi Sugiyama, Hayato Sugiyama, Chika Sato and Maki Morizumi

Abstract

We consider the application of microwave energy to a material. The effects of the electromagnetic field on the material and of the material on the electromagnetic field will be described, focusing on the dielectric relaxation phenomenon of the liquid. The dielectric permittivity of mixtures is discussed by extending Debye relaxation to explain how the material behaves with respect to an electric field. We will also consider the energy that the electric field imparts to the material, both thermally and nonthermally. We will develop this relation and describe what form it should take if there is a nonthermal effect in the chemical reaction field under microwave irradiation.

Keywords: microwave chemistry, complex permittivity, Debye relaxation, nonthermal effect, Arrhenius equation

1. Introduction

There has been much debate about whether microwave irradiation acts as heat in chemical synthesis or whether it has a nonthermal effect [1–6]. This problem has been discussed in many cases based on changes in the reaction rate and in the selectivity of the difference between the results with and without the application of microwaves. What is particularly important here is whether the temperatures of the two conditions to be compared are exactly the same. In microwave irradiation, it is difficult to use a general thermometer such as a metal thermocouple or an alcohol thermometer. This is because the distribution of the electromagnetic field changes significantly due to the insertion of a metal material (thermocouple or mercury thermometer), or because the indicator material (alcohol) itself is heated. As an alternative, an indirect method such as measuring the temperature from radiation on the surface of the vessel is used. In the comparison between microwave irradiation and non-irradiation, if the measurement does not correctly indicate the internal

IntechOpen

temperature, the difference may be due to the microwave irradiation condition being higher than the non-irradiation condition.

The heat source in microwave irradiation is the loss of electromagnetic wave energy, i.e., loss of the electric or the magnetic fields due to undulation of the molecule itself. Therefore, the movement behavior varies depending on the molecular species of the irradiated material. A different momentum obtained for each molecule means that it is not in a thermal equilibrium state, meaning that it does not match the definition of temperature, which requires an isotropic equilibrium motion.

The aforementioned is an inductive argument that discusses differences due to microwave irradiation, such as changes in reaction rates and in selectivity. A lot of data are reported every year, but the interpretations are diverse, and there are cases where it is "both hard to explain and hard to ignore" [1].

On the other hand, in this review, the original physical meaning is examined based on the dielectric relaxation phenomenon and how the material behaves under microwave irradiation. Based on this, we will discuss deductions about what action should be generated if there is an effect other than heat based on principles, rather than data.

2. Material properties (relaxation properties)

2.1 Permittivity and refractive index

When an object is heated by irradiation with microwaves, the microwave energy is attenuated inside the object [7]. The Beer–Lambert law in optics can also be applied in the microwave region. The refractive index, n, and the attenuation factor, k, can be combined as a complex refractive index n^*, as shown by Eq. (1):

$$n^* = n - jk, \tag{1}$$

where j is the square root of -1. When the wave has a cosine signal $s(t, x)$ as a function of time t and position x, the n and k correspond to the propagation and attenuation velocities in the phasor formula (Eq. (2)) as shown in **Figure 1**:

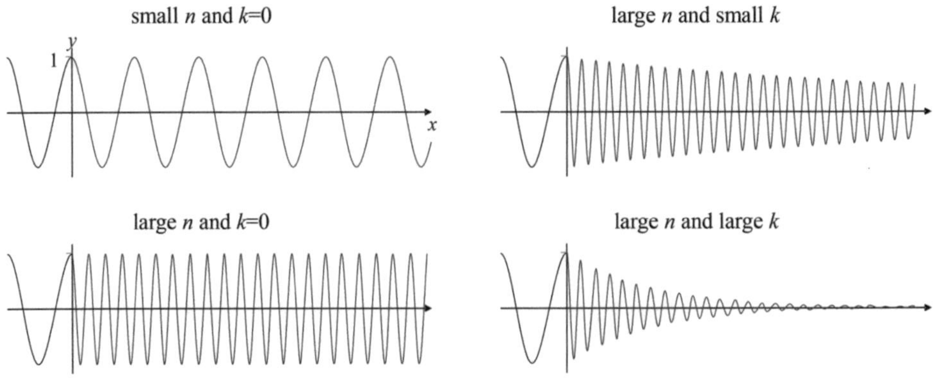

Figure 1.
Undulation and complex refractive index at t = 0.

DOI: http://dx.doi.org/10.5772/intechopen.106257

$$s(x,t) = A\cos(n^*x + \omega t) = \mathrm{Re}\left(Ae^{-j(n-jk)x}e^{-j\omega t}\right). \tag{2}$$

Physical properties in the microwave region are indicated by the complex permittivity ε^* [F/m] and complex permeability μ^* [H/m]. The meanings of these terms can be explained by definition of the values of basic physical constants. The speed of light propagation through the vacuum (c = 2.99792458 $\times$ 10^8 m/s) is defined value. After May 20, 2019, the magnetic constant μ_0 changed from the defined value ($4\pi \times 10^{-7}$ H/m) to the experimentally determined value (1.25663706212 (19) $\times$ 10^{-6} H/m) [8]. The electric constant ε_0 is derived from these constants (Eq. (3)):

$$\varepsilon_0 = \frac{1}{\mu_0 c^2}. \tag{3}$$

When Eq. (3) is transformed to Eq. (4), c represents the reciprocal of the square root of ε_0 and μ_0:

$$c = \frac{1}{\sqrt{\varepsilon_0\mu_0}}. \tag{4}$$

To be precise, vacuum is not a material, but electromagnetic waves propagate through it. Since the same relationship applies to materials, they can be treated equally well in terms of mathematical expressions. The velocity v [m/s] of the electromagnetic wave propagating through a material is the reciprocal of square root of the product of the material's permittivity ε [F/m] and permeability μ [H/m] (Eq. (5)):

$$v = \frac{1}{\sqrt{\varepsilon\mu}}. \tag{5}$$

Therefore, the refractive index n is obtained by Eq. (6):

$$n = \frac{c}{v} = \frac{\sqrt{\varepsilon\mu}}{\sqrt{\varepsilon_0\mu_0}} = \sqrt{\varepsilon_r\mu_r} \quad \text{where } \varepsilon = \varepsilon_0\varepsilon_r,\ \ \mu = \mu_0\mu_r. \tag{6}$$

Here, the relative permittivity ε_r [nd] and the relative permeability μ_r [nd] are coefficients based on ε_0 and μ_0, and n is dimensionless. In the following, the dimensionless value is expressed as [nd].

When attenuation of electromagnetic waves occurs in a material, the complex relative permittivity ε_r^* [nd] and the complex relative permeability μ_r^* [nd] are used. Therefore, the relationship with n^* is as follows (Eq. (7)):

$$n^* = \sqrt{\varepsilon_r^*\mu_r^*} \quad \text{where } \varepsilon_r^* = \varepsilon_r' - j\varepsilon_r'',\ \ \mu_r^* = \mu_r' - j\mu_r''. \tag{7}$$

The superscripts ‘and “indicate a real part and an imaginary part, respectively.

Discussions dealing only with dielectrics generally introduce important assumptions here. Since dielectrics often do not exhibit magnetism, the permeability is considered to be the same as that of vacuum, and $\mu_r{}^*$ is set to $1-j0$. Devices that measure permittivity (actually complex relative permittivity) often base their calculations on this assumption, so one should be careful when measuring materials with magnetism or high conductivity. Under this assumption, Eq. (7) is approximated by Eqs. (8) and (9):

$$n^* = \sqrt{\varepsilon_r^*}, \tag{8}$$

$$(n - jk)^2 = \varepsilon_r' - j\varepsilon_r''. \tag{9}$$

Here, Eq. (1) is reviewed again. Since n is the ratio of the propagation velocity in the material to that in the vacuum, it can be regarded as the ratio of the wavelength λ_0 [m] in the vacuum to the wavelength λ [m] in the material (Eq. (10)):

$$n = \frac{c}{v} = \frac{\lambda_0}{\lambda}. \tag{10}$$

On the other hand, since attenuation does not occur in a vacuum, it is difficult to understand k as a ratio. Therefore, a distance δ [m] at which an electric field intensity E [V/m] becomes $1/e$ = 36.8% is used. Here, e is the Napier number and ω [rad / s] is the angular frequency. δ is called the skin depth and has dimensions of length. As δ decreases, the amount of attenuation increases, indicating a large k (Eq. (11)):

$$k = \frac{c}{\omega\delta} = \frac{\lambda_0}{2\pi\delta}. \tag{11}$$

Furthermore, when Eq. (9) is transformed, Eqs. (12) and (13) are obtained:

$$n = \left[\frac{1}{2}\varepsilon_r'\left(\sqrt{1 + \left(\varepsilon_r''/\varepsilon_r'\right)^2} + 1\right)\right]^{1/2}, \tag{12}$$

$$k = \left[\frac{1}{2}\varepsilon_r'\left(\sqrt{1 + \left(\varepsilon_r''/\varepsilon_r'\right)^2} - 1\right)\right]^{1/2}. \tag{13}$$

From these equations, n and k are obtained from the ε_r^* of the material. When n is large, the microwave that has progressed is greatly refracted. In particular, a cylindrical vessel collects power at the center as in the case of a lens, and heating may proceed locally. **Figure 2** shows an example simulating the electromagnetic field distribution

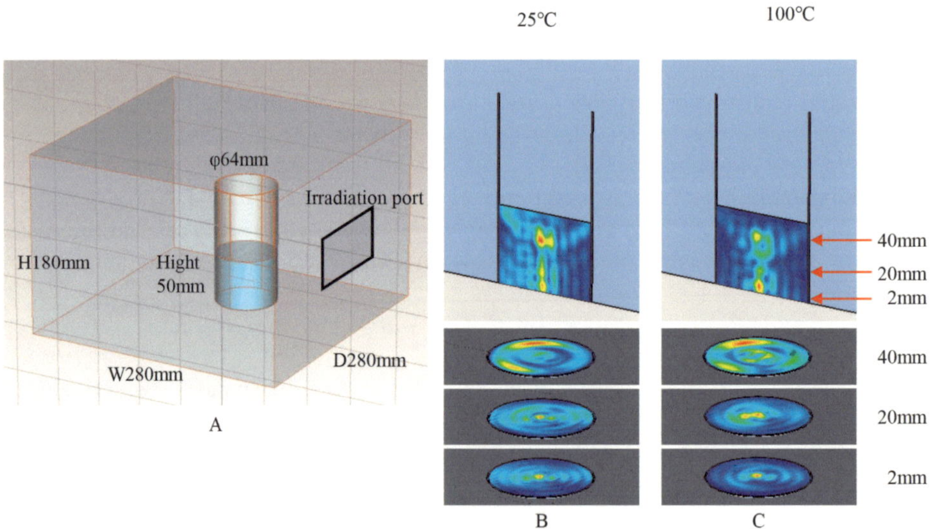

Figure 2.
Simulation of the electromagnetic field of water in a cylindrical vessel in an oven-type furnace. A: Oven shape. B: PLD of 25°C water. C: PLD of 100°C water.

of water in a cylindrical container. When the water temperature is uniform, the power loss density (PLD) is concentrated in the center [9].

2.2 Penetration depth and skin depth

When the same material is irradiated with electromagnetic waves having the same frequency, the applied power intensity P [W/m^3] is proportional to the square of the electric field intensity E [V/m]. When the attenuation is large, the microwave may not reach the deep part of the vessel. Although the skin depth, δ, has been described earlier, there is a penetration depth L [m] as a similar index [7]. The distance at which the power intensity P becomes 1/e due to the loss during propagation is expressed as $L_{1/e}$ [m]. In this case, penetration depth $L_{1/e}$ is half of the skin depth, δ. Microwaves are not absent beyond this depth.

There are two methods for describing $L_{1/e}$, as shown in Eqs. (14) and (15):

$$L_{1/e} = \frac{\lambda_0}{4\pi}\left[\frac{2}{\varepsilon_r'\left(\sqrt{1+\left(\varepsilon_r''/\varepsilon_r'\right)^2}-1\right)}\right]^{1/2}, \tag{14}$$

$$L_{1/e} = \frac{\lambda_0}{2\pi\sqrt{\varepsilon_r'}\tan\delta}. \tag{15}$$

The δ in Eq. (15) is not a skin depth, but a value indicating dielectric loss in a narrow definition as $\tan\delta$ (Eq. (16)):

$$\tan\delta = \frac{\varepsilon_r''}{\varepsilon_r'}. \tag{16}$$

Eq. (15) can be obtained from a modification in which the second and subsequent terms are ignored in the Maclaurin expansion when $\tan^2\delta \to 0$ in Eq. (14). Therefore, when using Eq. (15), it is assumed that $\tan\delta \to 0$. On the other hand, during microwave heating, the material of $\tan\delta \to 0$ does not heat, as will be described later. Therefore, if the target is not $\tan\delta \to 0$, it is necessary to pay attention to whether the value based on the latter formula deviates from the premise.

2.3 Plotting on bode and Nyquist diagrams

The correlation with the horizontal axis representing frequency and the vertical axis representing complex permittivity is called a Bode diagram. The Bode diagram of the water is shown in **Figure 3**, indicating that the dielectric constants ε_r' and ε_r'' are functions of the (angular) frequency [10]. This is represented by the Eq. (17):

$$\varepsilon = \varepsilon_0\varepsilon_r^*(\omega) = \varepsilon_0\left\{\varepsilon_r'(\omega) - j\varepsilon_r''(\omega)\right\}. \tag{17}$$

When the complex permittivity at each frequency is plotted on a Nyquist diagram with a real part on the horizontal axis and an imaginary part on the vertical axis, they draw a semicircular locus as shown in **Figure 3**. Such behavior is called Debye relaxation. Debye relaxation is a behavior commonly found in nonionic liquid materials.

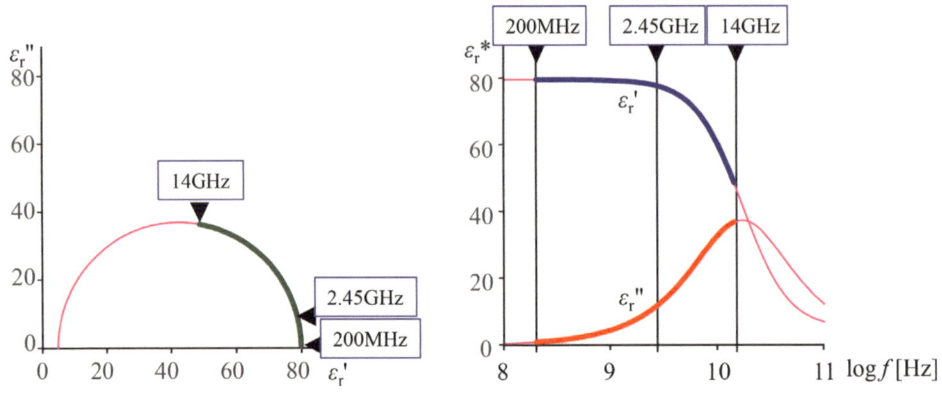

Figure 3.
Nyquist diagram and bode diagram of water (200 MHz–14 GHz).

Examples that cannot be applied include cases where the relaxation frequency is not single, and those where a conductive material is included (described at 2.10 and 2.11).

2.4 Relationship between the Debye relaxation formula and the bode/Nyquist diagram

In the previous section, we described how many liquids show a characteristically semicircular geometric locus due to Debye relaxation. We will return to the basics to explain why and what information can be gleaned below. The characteristic behavior in the microwave band is called dielectric relaxation. The deformation of electron clouds and molecular structures is a response in the UV and IR bands and is faster than in the microwave band. These contributions are prompt responses to undulated fields.

On the other hand, molecular orientation is a phenomenon based on rotation of an electric dipole. A large moment like a molecule causes a time delay in orientation with respect to field changes. The time delay referred to here is a phase delay and does not vibrate at a different period from that of the applied external field. Since the molecule cannot rotate if the external field vibration is too fast, but it can follow a too-slow external field vibration without time delay, the behavior is distributed around a specific vibration frequency [11–14]. The prompt response is only a propagation delay and does not contribute to the loss. This is expressed as $\varepsilon_r(\infty)$ only in the real part. At the current time t, the application of an external electric field $E(t)$ generates an electric flux density $D(t)$ in the material.

The part of the electric flux density in the material $D_p(t)$ (p: prompt) pertaining to prompt response is expressed by Eq. (18):

$$D_p(t) = \varepsilon_0 \varepsilon_r(\infty) E(t). \tag{18}$$

On the other hand, the contribution of the delayed response includes not only the electric field $E(t)$ at the current time t but also the influence of the electric field $E(u)$ at the previous time u. Therefore, the electric flux density $D_d(t)$ (d: delayed) of the dielectric following the delay is expressed by Eq. (19), integrated from start time 0 to the current time t:

$$D_d(t) = \int_0^t E(u) f(t-u) du. \tag{19}$$

In Eq. (19), $f\ (t\text{-}u)$ represents a response function in terms of the previous time u and the current time t. From Eqs. (18) and (19), the electric flux density $D\ (t)$ of the dielectric at the current time t is expressed by Eq. (20):

$$D(t) = \varepsilon_0 \varepsilon_r(\infty) E(t) + \int_0^t E(u) f(t-u) du. \tag{20}$$

$D(t)$ in Eq. (20) can also be expressed as the product of the applied electric field $E(t)$ and the dielectric constant $\varepsilon^*(\omega)$ of the material. Therefore, Eq. (21) can be obtained:

$$D(t) = \varepsilon_0 \varepsilon_r^*(\omega) E(t). \tag{21}$$

Substituting Eq. (21) into Eq. (20) and rewriting the response function to $f\ (x)$ yields Eq. (22):

$$\varepsilon_0 \left\{ \varepsilon_r^*(\omega) - \varepsilon_r(\infty) \right\} = \int_0^\infty e^{-j\omega x} f(x) dx. \tag{22}$$

Next, an appropriate expression is set for a response function. In Debye-type relaxation, Eq. (23) is used as a response function to Eq. (7):

$$f(x) = \varepsilon_0 \{ \varepsilon_r(0) - \varepsilon_r(\infty) \} \frac{e^{-x/\tau}}{\tau}. \tag{23}$$

Here, τ [s/rad] is the relaxation time, which is the reciprocal of the angular frequency ω_0 [rad/s] at which the vibration phase is delayed by $\pi/2$ (90 degrees). A very slowly undulating field is effectively the same as a static field. Therefore, the complex relative dielectric constant is also only the real part, and this is represented by $\varepsilon_r(0)$. As shown in the Nyquist diagram of **Figure 3**, $\varepsilon_r'(\omega)$, which is the real part of $\varepsilon_r^*(\omega)$, is between $\varepsilon_r(0)$ and $\varepsilon_r(\infty)$. Therefore, $f\ (x)$ can be interpreted as having a proportional coefficient of $\varepsilon_r(0)\text{-}\varepsilon_r(\infty)$, which is the difference between the real parts.

When Eq. (23) is substituted into Eq. (22) and transformed, $\varepsilon_r^*(\omega)$ is expressed by Eqs. (24)–(26). These are Debye's dispersion equations:

$$\varepsilon_r^*(\omega) = \varepsilon_r(\infty) + \frac{\varepsilon_r(0) - \varepsilon_r(\infty)}{1 + j\omega\tau}; \tag{24}$$

$$\varepsilon_r'(\omega) = \operatorname{Re} \varepsilon_r^*(\omega) = \varepsilon_r(\infty) + \frac{\varepsilon_r(0) - \varepsilon_r(\infty)}{1 + \omega^2\tau^2}; \tag{25}$$

$$\varepsilon_r''(\omega) = \operatorname{Im} \varepsilon_r^*(\omega) = \frac{\{\varepsilon_r(0) - \varepsilon_r(\infty)\}\omega\tau}{1 + \omega^2\tau^2}. \tag{26}$$

Eliminating ω_τ from Eqs. (25) and (26) leads to the relationship of Eq. (27). Therefore, when $\varepsilon_r'(\omega)$ is on the horizontal axis and $\varepsilon_r''(\omega)$ is on the vertical axis, a semicircle is drawn as shown in **Figure 3**:

$$\left\{ \varepsilon_r'(\omega) - \frac{\varepsilon_r(0) + \varepsilon_r(\infty)}{2} \right\}^2 + \varepsilon_r''(\omega)^2 = \left\{ \frac{\varepsilon_r(0) - \varepsilon_r(\infty)}{2} \right\}^2. \tag{27}$$

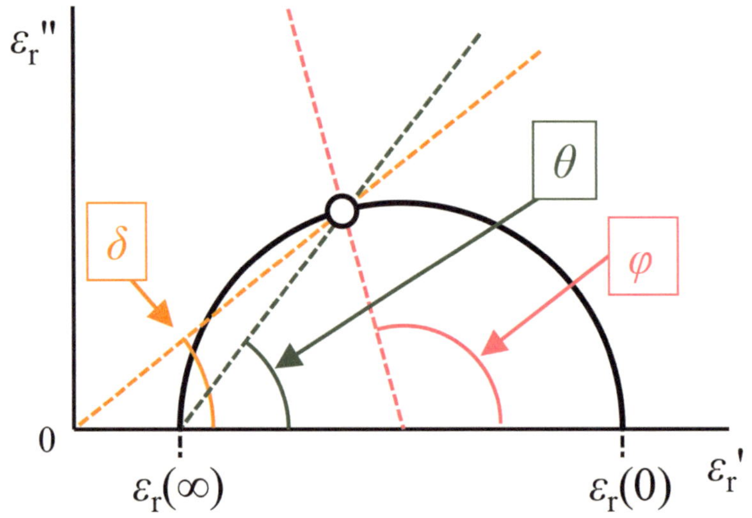

Figure 4.
Three angles on a semicircle.

From the aforementioned, if the measured values are plotted on a semicircle, it indicates that the material has a response that can be explained by Debye relaxation theory and is not in a special state.

2.5 Three angles on the semicircle: φ, θ, and δ

The dielectric loss is also written as tan δ. This is caused by the phase delay angle δ of the voltage and current when an alternating electric field is applied to the dielectric; it is defined by Eq. (16). The complex dielectric constant is calculated by measuring the loss and the propagation delay. The physical meaning of this value is a tangent, where δ is the angle between the horizontal axis and the line segment from the origin to the circumference (**Figure 4**).

Substituting Eqs. (25) and (26) into Eq. (16) shows that tan δ is a function of ω:

$$\tan\delta = \frac{\{\varepsilon_r(0) - \varepsilon_r(\infty)\}\omega\tau}{\varepsilon_r(0) + \varepsilon_r(\infty)\omega^2\tau^2}. \tag{28}$$

If the central angle φ is determined at an arbitrary point on the semicircle when $\varepsilon_r(0)$ is $\varphi = 0°$, $\varepsilon_r(\infty)$ shows $\varphi = 180°$ [15]. Therefore, φ indicates a phase delay of molecular motion in the delayed response. The tan φ at an arbitrary frequency is obtained geometrically by the following equation:

$$\tan\varphi = \frac{\varepsilon_r''(\omega)}{\varepsilon_r'(\omega) - \frac{\varepsilon_r(0)+\varepsilon_r(\infty)}{2}}. \tag{29}$$

Substituting Eqs. (25) and (26) into Eq. (29) eliminates $\varepsilon_r(0)$ and $\varepsilon_r(\infty)$, as shown in Eq. (30). Therefore, the influence of φ upon temperature change means that it is based only on the change of the dielectric relaxation time τ when the frequency is constant:

$$\tan\varphi = \frac{2\omega\tau}{1-\omega^2\tau^2} = -\frac{1}{\sinh(\ln(\omega\tau))} = -\mathrm{csch}(\ln(\omega\tau)). \tag{30}$$

As shown in **Figure 4**, if θ is defined as an angle formed by a straight line extending from the left intersection of the semicircle and the horizontal axis toward the measurement point and the horizontal axis, then θ is a circumferential angle of φ. Here, tan θ is expressed by Eq. (31):

$$\tan\theta = \frac{\varepsilon_r''(\omega)}{\varepsilon_r'(\omega) - \varepsilon_r'(\infty)}. \tag{31}$$

Substituting Eqs. (25) and (26) into Eq. (31) yields Eq. (32):

$$\tan\theta = \omega\tau; \tag{32}$$

transforming it yields Eq. (33) for relaxation time τ [s/rad]:

$$\tau = \frac{1}{\omega}\tan\theta = \frac{1}{\omega}\frac{\varepsilon_r''(\omega)}{\varepsilon_r'(\omega) - \varepsilon_r(\infty)}. \tag{33}$$

When calculating τ, in a region far from θ = 45°(φ = 90°), even if ω changes logarithmically, φ does not change significantly, and the calculation of τ has a large error. On the other hand, in the vicinity of φ = 90°, $\varepsilon_r^*(\omega)$ changes sharply with respect to ω. Therefore, when calculating τ using Eq. (31), it is preferable to perform evaluation with a measurement point in the vicinity of $2\theta = \varphi$ = 90°.

From Eq. (32), $\varepsilon_r''(\omega)$ is maximum at $\omega = 1/\tau$. In actual measurements, the function to be swept is often denoted by f[Hz] instead of ω[rad/s]. It should be noted that in many references, τ is sometimes written in terms of the reciprocal of the relaxation frequency f_c which is defined in Eq. (58). In this case, the unit of the derived τ is [s], which is 2π times τ[s/rad]. In **Table 1**, the unit of τ is written in [s] instead of [s/rad].

Entry		**f_c [GHz]**	**τ [ps]**	**$\varepsilon_r(0)$**	**$\varepsilon_r(\infty)$**	**ε_r*_(2.45 GHz)**		**ε_r*_(5.8 GHz)**	
						Calcd.	**Found**	**Calcd.**	**Found**
1	H_2O	16.8	59.4	79.8	4.8	78.2–*j*10.7	78.9–*j*10.8	71.8–*j*23.1	72.0–*j*23.3
2	ProC	3.5	287	66.2	7.4	46.7–*j*27.6	46.5–*j*27.8	23.0–*j*26.0	22.9–*j*26.2
3	DMSO	7.7	130	47.6	7.9	44.0–*j*11.5	44.2–*j*11.1	33.2–*j*19.1	33.6–*j*19.0
4	DMAc	8.6	116	40.5	6.2	37.9–*j*9.0	37.8–*j*9.2	29.8–*j*15.9	29.7–*j*16.1
5	$MeNO_2$	32.3	31.0	37.1	11.0	37.0–*j*2.0	37.4–*j*2.3	36.3–*j*4.5	36.4–*j*5.1
6	DMF	13.9	72.0	37.6	8.2	36.7–*j*5.0	36.9–*j*5.2	33.2–*j*10.5	33.0–*j*10.8
7	MeCN	39.2	25.5	35.6	12.2	35.5–*j*1.5	35.4–*j*1.8	35.1–*j*3.4	35.0–*j*4.4
8	Me-Im	5.0	198	38.0	5.5	31.8–*j*12.8	31.8–*j*12.6	19.5–*j*16.1	19.7–*j*16.2
9	NMP	6.8	148	33.4	6.0	30.2–*j*8.8	30.2–*j*8.7	21.8–*j*13.5	22.0–*j*13.7
10	$PhNO_2$	3.5	285	35.0	4.5	25.0–*j*14.3	24.6–*j*14.3	12.7–*j*13.5	12.2–*j*13.4

Entry		f_c [GHz]	τ [ps]	$\varepsilon_r(0)$	$\varepsilon_r(\infty)$	ε_r*_(2.45 GHz)		ε_r*_(5.8 GHz)	
						Calcd.	**Found**	**Calcd.**	**Found**
11	MeOH	3.0	338	33.7	6.1	22.5–*j*13.6	22.2–*j*13.2	11.8–*j*11.2	11.6–*j*11.1
12	PhCN	4.5	223	26.0	4.4	21.1–*j*9.1	21.4–*j*8.7	12.5–*j*10.5	12.7–*j*10.6
13	Me_2CO	44.4	22.5	21.2	5.8	21.2–*j*0.8	21.0–*j*1.0	21.0–*j*2.0	21.1–*j*2.4
14	MEK	24.2	41.3	18.7	9.0	18.6–*j*1.0	18.6–*j*1.1	18.1–*j*2.2	18.0–*j*2.4
15	PhCHO	5.2	191	18.9	4.0	16.2–*j*5.7	16.2–*j*5.7	10.7–*j*7.4	10.8–*j*7.5
16	DCE	15.4	65.1	10.4	5.2	10.2–*j*0.8	10.2–*j*0.9	9.7–*j*1.7	9.7–*j*1.9
17	CH_2Cl_2	—	—	9.3	8.5	—	9.3–*j*0.1	—	9.1–*j*0.5
18	EtOH	0.9	1100	25.8	4.6	7.2–*j*7.0	7.5–*j*7.0	5.1–*j*3.3	5.3–*j*3.4
19	MeI	17.9	55.9	7.3	5.7	7.3–*j*0.2	7.2–*j*0.2	7.2–*j*0.5	7.1–*j*0.6
20	AcOEt	13.0	76.6	6.3	4.6	6.2–*j*0.3	6.2–*j*0.3	6.0–*j*0.6	6.0–*j*0.7
21	PhCl	10.4	96.3	5.8	3.2	5.7–*j*0.6	5.7–*j*0.6	5.2–*j*1.1	5.2–*j*1.1
22	PhF	13.8	72.6	5.8	4.0	5.7–*j*0.3	5.7–*j*0.3	5.5–*j*0.6	5.5–*j*0.6
23	PhBr	7.5	134	5.8	3.1	5.6–*j*0.8	5.5–*j*0.7	4.8–*j*1.3	4.9–*j*1.3
24	$CHCl_3$	11.1	89.9	5.1	3.9	5.0–*j*0.2	5.1–*j*0.0	4.8–*j*0.5	4.9–*j*0.4
25	PhI	5.2	192	5.0	3.0	4.7–*j*0.8	4.7–*j*0.7	3.9–*j*1.0	3.9–*j*1.0
26	Et_2O	—	—	4.5	4.1	—	4.5–*j*0.1	—	4.4–*j*0.2
27	2-PrOH	0.4	2730	20.1	3.4	3.8–*j*2.4	4.1–*j*2.7	3.5–*j*1.0	3.6–*j*1.3
28	CPME	27.8	36.0	4.0	2.3	4.0–*j*0.1	3.9–*j*0.3	3.9–*j*0.3	3.9–*j*0.4
29	1-BuOH	0.3	3170	16.3	3.4	3.6–*j*1.6	3.9–*j*2.0	3.4–*j*0.7	3.5–*j*1.1
30	2-BuOH	0.3	3540	15.4	3.2	3.3–*j*1.4	3.6–*j*1.6	3.2–*j*0.6	3.3–*j*0.9
31	*i*-BuOH	0.2	4070	16.1	3.1	3.3–*j*1.3	3.5–*j*1.6	3.2–*j*0.5	3.2–*j*0.9
32	*t*-BuOH	0.3	2910	11.6	2.9	3.1–*j*1.2	3.3–*j*1.5	2.9–*j*0.5	3.0–*j*0.8
33	PhMe	—	—	2.5	2.4	—	2.5–*j*0.1	—	2.4–*j*0.1
34	PhH	—	—	2.4	2.3	—	2.4–*j*0.1	—	2.4–*j*0.0
35	*c*-Hex	—	—	2.2	2.1	—	2.1–*j*0.1	—	2.1–*j*0.0
36	*n*-Hex	—	—	2.0	1.4	—	2.0–*j*0.0	—	2.0–*j*0.0

Entry 1: distilled water; 2: propylene carbonate; 3: dimethyl sulfoxide; 4: N,N-dimethylacetamide; 5: nitromethane; 6: N,N-dimethylformamide; 7: acetonitrile; 8: N-methylimidazole; 9: N- methyl-2-pyrrolidone; 10: nitrobenzene; 11: methanol; 12: benzonitrile; 13: acetone; 14: methyl ethyl ketone; 15: benzaldehyde; 16: 1,2-dichloroethane; 17: dichloromethane; 18: ethanol; 19: methyl iodide; 20: ethyl acetate; 21: chlorobenzene; 22: fluorobenzene; 23: bromobenzene; 24: chloroform; 25: iodobenzene; 26: diethyl ether; 27: 2-propanol; 28: cyclopentyl methyl ether; 29: 1-butanol; 30: 2-butanol; 31: isobutyl alcohol; 32: tert*-butyl alcohol; 33: toluene; 34: benzene; 35: cyclohexane; 36:* n-*hexane.* f_c*: relaxation frequency obtained by fitting,* τ*: relaxation time, calcd.* ε_r (f)*: calculated* ε_r **, found: measured* ε_r **.*

Table 1.
Physical property values of several liquids at room temperature [16].

2.6 Loss calculation formula

The propagation of the electromagnetic wave energy of the microwave is propagation of electromagnetic field vibration. Specifically, this vibration is caused by continuously and repeatedly converting the electric field energy into magnetic field energy and vice versa. The propagation equations are shown as follows (Eqs. (34)–(38)):

$$\mathrm{rot}\mathbf{H} = \mathbf{i} + \frac{\partial \mathbf{D}}{\partial t}; \tag{34}$$

$$\mathbf{B} = \mu \mathbf{H}; \tag{35}$$

$$\mathrm{rot}\mathbf{E} = -\frac{\partial \mathbf{B}}{\partial t}; \tag{36}$$

$$\mathbf{D} = \varepsilon \mathbf{E}; \tag{37}$$

$$\mathbf{i} = \sigma \mathbf{E}. \tag{38}$$

Here, **E**, **H**, **D**, **B**, and **i** are the electric field intensity [V/m], magnetic field intensity [A/m], electric flux density [C/m^2], magnetic flux density [Wb/m^2], and current density [A/m^2], all of which are vector quantities. In addition, the physical property coefficient of the material is transferred as a permittivity ε [F/m], a conductivity σ [S/m], and a permeability μ [H/m]. The loss of propagation energy means that the vector quantity has been lost when converted to the other field. This loss is mainly regarded as a conversion to heat. The loss equation calculated based on the physical property values is derived below [11, 12, 17].

The applied E is expressed by using a phasor as follows:

$$E = E_0 e^{j\omega t}. \tag{39}$$

D in the dielectric produced by aligning the directions of the dielectric molecules undulates with a phase shift (delay) of δ. δ is a value expressed by Eq. (16) and is not φ, which is an undulation phase delay of the molecule:

$$D = D_0 e^{j(\omega t - \delta)}. \tag{40}$$

From Eq. (37), the product of E, ε_0, and $\varepsilon_r{}^*$ gives D. Therefore, the phase delay of D can also be expressed by the complex permittivity (Eq. (41)):

$$D = \varepsilon_0 \varepsilon_r^* E. \tag{41}$$

Substituting Eqs. (39) and (40) into Eq. (41) and converting the exponential function to a trigonometric function using Euler's formula yields the Eqs. (42) and (43). Eq. (42) divided by Eq. (43) matches Eq. (16):

$$D_0 \cos \delta = \varepsilon_0 \varepsilon_r' E_0; \tag{42}$$

$$D_0 \sin \delta = \varepsilon_0 \varepsilon_r'' E_0. \tag{43}$$

Next, the energy consumption per unit volume when the electric flux density of the dielectric changes by dD is determined as dU. This dU is obtained by the product of E and dD. When dD is integrated over one period of vibration ($1/f = 2\pi/\omega$), the energy consumed by the dielectric during one period is obtained as w,

$$w = \int_0^{2\pi/\omega} dU = \int_0^{2\pi/\omega} E dD = \int_0^{2\pi/\omega} E \frac{dD}{dt} dt. \tag{44}$$

In Eq. (44), E and D are the real parts:

$$E = E_0 \cos \omega t; \tag{45}$$

$$D = D_0 \cos (\omega t - \delta). \tag{46}$$

After transforming Eq. (46) with the cosine difference formula, differentiating with t gives Eq. (47):

$$\frac{dD}{dt} = -\omega \varepsilon_0 \varepsilon_r' E_0 \sin \omega t + \omega \varepsilon_0 \varepsilon_r'' E_0 \cos \omega t. \tag{47}$$

By substituting Eq. (47) into Eq. (44), we obtain Eq. (48):

$$w = \pi \varepsilon_0 \varepsilon_r'' E_0^2. \tag{48}$$

Since w is repeated f times per second (= $\omega/2\pi$ times), the energy W received by the dielectric from the electric field per unit volume / unit time is expressed by Eq. (49):

$$W = \frac{\omega}{2\pi} w = \frac{1}{2} \omega \varepsilon_0 \varepsilon_r'' E_0^2 = \pi f \varepsilon_0 \varepsilon_r'' E_0^2. \tag{49}$$

Rewriting Eq. (49) yields Eq. (50). Here, when E_0 is rewritten to $|\mathbf{E}|$, the electric field loss equation is obtained. In this paper, the dielectric loss based on ε obtained by Eq. (50) is defined as P_{ε_loss} [W/m^3]:

$$P_{\varepsilon\_loss} = \frac{1}{2} \omega \varepsilon_0 \varepsilon_r'' |E|^2. \tag{50}$$

When the frequency is very low and the change in the electric field is very slow, the molecules align their dipole moments in a direction that cancels the electric field without delay in proportion to the electric field strength, and this flux density can follow without delay. Since $\varphi \to 0$ and $\delta \to 0$, $\varepsilon_r'' \to 0$ from Eq. (43), $P_{\varepsilon\_loss} \to 0$ from Eq. (50), and the electromagnetic wave energy loss is small. On the other hand, if the frequency is very high and the change in the electric field is very fast, the movement of the molecules cannot respond to the alternating electric field, and the application direction reverses before aligning the dipole moments. As a result, since $\varphi \to \pi$ and $\delta \to 0$, $\varepsilon_r'' \to 0$, so $P_{\varepsilon\_loss} \to 0$ and the electromagnetic wave energy loss is small.

2.7 Difference between tan δ and ε_r'' in loss

If the target frequencies are the same, ω may be regarded as a constant. According to the aforementioned equation, the loss appears to be proportional to ε_r'', but the actual loss is not determined solely by the difference in ε_r''. The meaning of Eq. 50 indicates that if E is constant, the amount of power loss per unit volume is proportional to ε_r''. However, the propagation speed of electromagnetic waves decreases with the refractive index n, which varies with ε (and μ).

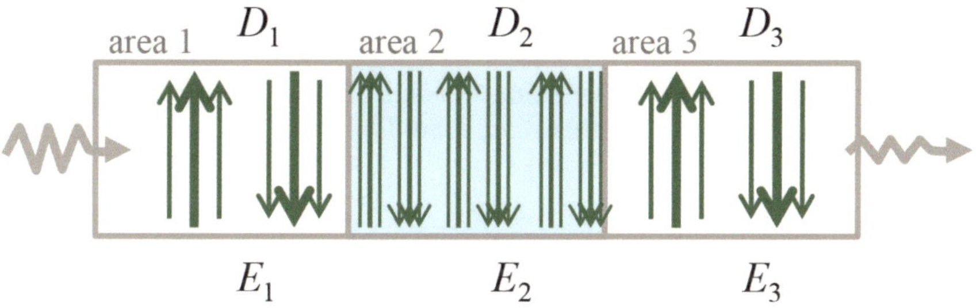

Figure 5.
Model structure of heating considerations. Area 1:Air, area 2:Water, area 3:Air.

Figure 5 shows a model in which microwaves pass through media in the order of air, water, and air. Here, for the sake of simplicity, the electric flux is indicated by a straight arrow, and the reflected wave at the boundary is not considered. Wavelength reduction is considered first [18]. The electric flux density in air is D_1 [C/m^2], and that in water is D_2 [C/m^2]. Since wavelength shortening occurs in water with a large ε_r', the interval (density) of arrows in D_2 increases to $(\varepsilon_r)^{0.5}$ times in D_1 according to Eq. (8). Next, the electric field strength is considered. The electric field strength in air is E_1 [V/m], and that in water is E_2 [V/m]. In water where ε_r' is large, the dipole of water cancels the applied electric field so that the intensity decreases to $1/\varepsilon_r$ according to the constitutive Eq. (37). Combining these two effects, the electric field strength E_2 in water attenuates to $1 / (\varepsilon_r)^{0.5}$ times the electric field strength E_1 in air as shown in Eq. (51):

$$\begin{cases} D_2 = \sqrt{\varepsilon_r'} D_1 \\ D_1 = \varepsilon_0 E_1 \\ D_2 = \varepsilon_0 \varepsilon_r' E_2 \end{cases} \quad \text{therefore } E_2 = \frac{1}{\sqrt{\varepsilon_r'}} E_1. \tag{51}$$

In actual examinations, it is difficult to measure the electric field strength inside the irradiation target. Therefore, irradiation with a predetermined irradiation power is performed. Thus, the following interpretation is derived:

a. Eq. (52) indicates that the loss is proportional to ε_r'' when the applied electric field is constant:

$$P_{\varepsilon\_loss} = \frac{1}{2}\omega\varepsilon_0\varepsilon_r''|E_2|^2. \tag{52}$$

b. Eq. (53) indicates that the loss is proportional to $\tan\delta$ when the applied power is constant:

$$P_{\varepsilon\_loss} = \frac{1}{2}\omega\varepsilon_0\varepsilon_r''\left|\frac{E_1}{\sqrt{\varepsilon_r'}}\right|^2 = \frac{1}{2}\omega\varepsilon_0 \tan\delta |E_1|^2. \tag{53}$$

The applied electric field and applied power referred to here are the net electric field and power applied to the materials. As will be described later, not all irradiation power is always applied. The irradiated and reflected powers can be measured in area 1. The passing through power can be measured in area 3.

2.8 Maximum of tanδ

From the Nyquist diagram, the maximum value of tan δ is the tangent point through the origin. Therefore, it is obtained from Eq. (54):

$$\tan\delta_{\max} = \frac{\varepsilon_r(0) - \varepsilon_r(\infty)}{2\sqrt{\varepsilon_r(0)\varepsilon_r(\infty)}}. \tag{54}$$

Furthermore, $\varepsilon_r'(\omega)$ and $\varepsilon_r''(\omega)$ when tan δ is maximal are represented by Eqs. (55) and (56):

$$\varepsilon'_{r\,\tan\delta\max} = \frac{2\varepsilon_r(0)\varepsilon_r(\infty)}{\varepsilon_r(0) + \varepsilon_r(\infty)}, \tag{55}$$

$$\varepsilon''_{r\,\tan\delta\max} = \frac{\varepsilon_r(0) - \varepsilon_r(\infty)}{\varepsilon_r(0) + \varepsilon_r(\infty)}\sqrt{\varepsilon_r(0)\varepsilon_r(\infty)}. \tag{56}$$

The tan θ when tan δ becomes maximum is defined as tan $\theta_{\tan\delta\max}$. This is geometrically determined from **Figure 4**. The angular frequency $\omega_{\tan\delta\max}$ at this time is given by Eq. (32). When both are combined, Eq. (57) is derived:

$$\tan\theta_{\tan\delta\max} = \omega_{\tan\delta\max}\tau = \sqrt{\frac{\varepsilon_r(0)}{\varepsilon_r(\infty)}}. \tag{57}$$

The frequency at which ε_r'' is maximized is defined as f_c. Since tanθ is 1 at f_c, $2\pi f_c\,\tau$ is equal to 1 from Eq. (32) and **Figure 4**. Therefore, Eq. (57) becomes Eq. (58) by $f_{\tan\delta\max}$ and f_c:

$$f_{\tan\delta\max} = \sqrt{\frac{\varepsilon_r(0)}{\varepsilon_r(\infty)}}f_c. \tag{58}$$

The semicircle in the Nyquist diagram shows that the maximum values of ε_r'' and tan δ do not match. For example, water has a maximum value of ε_r'' at 18 to 22 GHz, but the maximum value of tan δ is on the higher frequency side. Since 2.45 GHz is much smaller than these, it appears to be less efficient at heating water. However, when the amount of absorption is large, attenuation occurs rapidly in the surface and does not penetrate into the inner side. Therefore, it cannot be said that a larger loss is always effective for heating the inside to a wide area.

2.9 Changes in ε_r^* with temperature

The complex dielectric constant according to Debye relaxation can be generalized by obtaining $\varepsilon_r(0)$, $\varepsilon_r(\infty)$, and τ by measuring in a wideband and fitting to a semicircle. From this relationship, it can be seen that the temperature, frequency, and complex permittivity have the following relationship:

1. When the temperature rises, the molecule becomes disturbed. As a result, the external response amount $\varepsilon_r'(\omega)$ decreases. This also applies to $\varepsilon_r(0)$.

2. When the temperature rises, the intermolecular bond becomes weaker and the relaxation time τ becomes shorter. As a result, the peak value of $\varepsilon_r''(\omega)$ shifts to the high frequency side.

3. If the dielectric loss is based on the response-phase difference φ, $\varepsilon_r'(\omega)$ and $\varepsilon_r''(\omega)$ are linked by the Kramers-Kronig relations and are not independent.

As shown in **Figure 6**, the Bode diagrams of $\varepsilon_r'(\omega)$ and $\varepsilon_r''(\omega)$ shift from $1 \rightarrow 2 \rightarrow 3$ with temperature rising [19]. Therefore, in the Nyquist diagram, the central angle φ indicated by the irradiation frequency f_i can be classified into six types depending on the position [11–13].

Type I is the case where the relaxation frequency f_c of the irradiated material before heating is much larger than f_i, and the φ of the irradiated material before heating is in the vicinity of 0π/8 to 2π/8. As the temperature rises, φ approaches 0 and the amount of loss decreases.

Type II is the case where f_c is slightly larger than f_i, and φ indicates 2π/8 to 4π/8. Since $\varepsilon_r''(\omega)$ is large, the temperature rises rapidly. However, as the temperature becomes high, φ decreases and the temperature rise rate greatly decreases.

Type III is when f_i is close to or slightly smaller than f_c, and φ is in the vicinity of 4π/8 to 5π/8. Since $\varepsilon_r''(\omega)$ is large as a whole, the temperature rise rate is high and $\varepsilon_r''(\omega)$ rises until φ becomes π/2 and then falls.

Type IV is when f_c is smaller than f_i, and φ indicates 5π/8 to 7π/8. The tan δ is a large region, and as temperature rises, φ approaches 4π/8, so $\varepsilon_r''(\omega)$ further increases. Thermal runaway due to uneven heating may occur.

Type V is the case where f_c is much smaller than f_i, and φ indicates 7π/8 to 8π/8. Although it has the property that the temperature rises due to irradiation and the amount of absorption increases, there is a case where the irradiation frequency is too high, such that there is a lot of transmission and heating is not sufficient.

Type 0 is not shown in this figure. This corresponds to the case where contributions to the original dielectric loss, such as that from nonpolar molecules, are very small.

When the irradiated f_i is considerably smaller than f_c, the change in $\varepsilon_r'(\omega)$ and $\varepsilon_r''(\omega)$ due to temperature rising is classified as type I, but when it irradiates a considerably larger f_c, it becomes a V type. Therefore, this classification also means that the classification changes depending on the irradiation frequency, even for the same material. As an example, the physical property values of several liquids and the value of τ calculated from a semicircle are shown in **Table 1**.

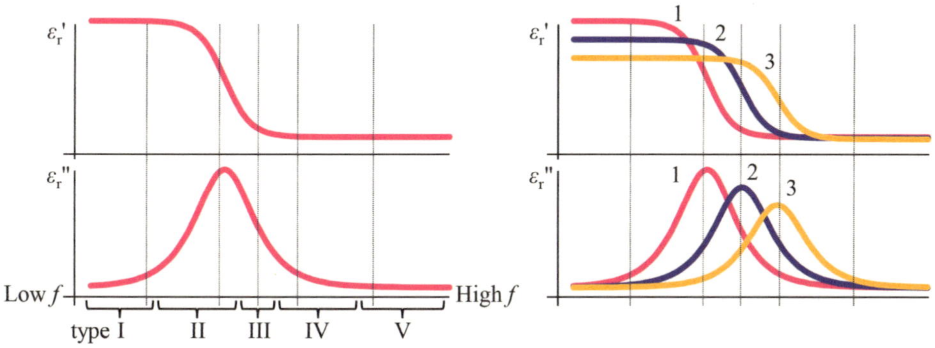

Figure 6.
Area division of the bode diagram. Left: area division; Right: temperature rising.

2.10 Liquid mixture

For the relaxation time τ in the system to show only one value, it is necessary for the entire material to be in a uniform state. This occurs when there is only one kind of electric dipole that controls the dielectric constant, and the surrounding molecules that control the relaxation time are also uniform. This means that τ is distributed according to the δ function [20]. In the mixed liquid, there are various types of molecules exhibiting dielectric loss and various types of surrounding molecules. Therefore, the relaxation time is not always one. In the model, when the $\varepsilon_r{}^*(\omega)$ of a system having the same $\varepsilon_r(0)$ and $\varepsilon_r(\infty)$, different τ is calculated. **Figure 7** shows that wideband complex permittivity plot is separated into two semicircles when the difference between the two τ is large. On the other hand, when the difference in τ is not large, the distortion of the semicircle is small. In the case of simple two-component mixing, the trajectory shown in the Nyquist diagram is considered to be similar to any in **Figure** 7. However, there is little distortion in the measurement range as shown in **Table 1**, and there is almost a single semicircle. This can be regarded as a response in which $\varepsilon_r(0)$, $\varepsilon_r(\infty)$, and τ show one average value. Furthermore, when τ is evaluated with respect to the mixing ratio, it continuously changes according to the composition, but it is not always shown on the straight line in terms of arithmetic mean. However, there is also a behavior that protrudes upward and downward.

Figure 8 shows the Argan diagram as a complex relative permittivity at 2.45 GHz by the reflection probe method [21–23]. One line indicates the nine mixtures made with a volume ratio of 9:1 to 1:9 between two pure liquids. If the $\varepsilon_r{}^*$ obtained in the mixed sample obeys the additive property, the connection should be a straight line, but in many cases a curve is shown. This is because $\varepsilon_r(0)$, $\varepsilon_r(\infty)$, and τ change due to the mixing of the two materials, such that $\varepsilon_r{}'$ and $\varepsilon_r{}''$ do not always attain a single arithmetic average value.

A plot of the responses at other frequencies on the Argan diagram is also shown in **Figure 8**. Apparently, these figures have changed greatly. However, as shown in point A, ethanol:water = 4:6 mixed solution, acetonitrile:propylene carbonate = 2:8 mixed solution, and acetone:propylene carbonate = 2:8 mixed solution had close $\varepsilon_r{}^*$ values, regardless of frequency. Similar intersection points were also observed in Group B (acetone:propylene carbonate = 9:1, cyclopentyl methyl ether:acetonitrile = 3:7) and Group C (ethanol:propylene carbonate = 7:3, methanol = 10). This means that if the average $\varepsilon_r(0)$, $\varepsilon_r(\infty)$, and τ obtained by mixing are close, $\varepsilon_r{}^*(\omega)$ matches, and therefore the same $\varepsilon_r{}^*$ is shown at different frequencies. Thus, when the liquid mixture characteristic is shown as a "train map," the intersection corresponding to the "transfer station" does not change even if the frequency does.

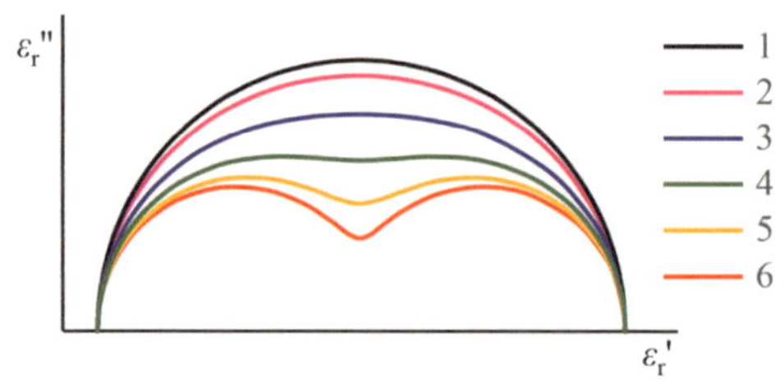

Figure 7.
Nyquist diagrams with two τ. Line 1: zero difference in τ, Line 2: small difference in τ ~ Line 6: large difference in τ.

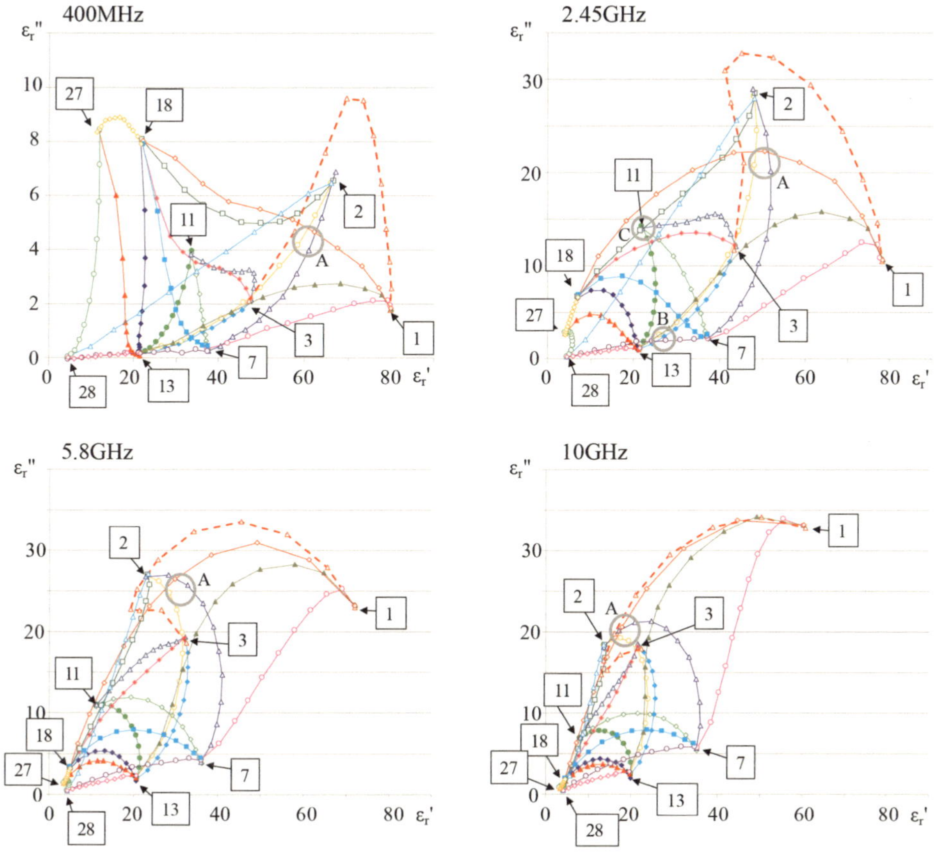

Figure 8.
Argan diagrams of liquid mixtures. 1: Water; 2: Propylene carbonate; 3: Dimethyl sulfoxide; 7: Acetonitrile; 11: Methanol; 13: Acetone; 18: Ethanol; 27: 2-propanol; 28: Cyclopentyl methyl ether.

2.11 Involvement of conductivity σ

The conductive material is heated by the conductive loss [10, 24–27]. This property agrees with the dielectric loss in that it is proportional to the square of the electric field strength. These are losses to the electric field and not to the magnetic field. Since ω is not included in the conduction loss equation, it occurs even when the frequency is zero. Conduction loss is a phenomenon in which materials with a single charge, that is, positive and negative ion atoms (or molecules), are accelerated in opposite directions by application of an electric field. An increase in the distance between the counterions means that the electrostatic potential of the material is increased. Also, if ions that should linearly move with constant acceleration are decelerated to constant speed, this means that resistance has made ions motion isotropic, or the surrounding molecules have received the kinetic energy of ions. This means that the current has been converted to Joule heat. It is clear that such a conduction loss differs from dielectric loss in which the charge distance in the molecule is constant. Therefore, in the partial dielectric including conductivity, the conductive and dielectric losses appear separately. For example, the Nyquist diagrams of 0.1-mol/L = NaCl aqueous solution are shown in **Figure 9**. Comparing to **Figure 3**, it is shown that the locus is changed by adding NaCl due to conductive loss.

In many cases, only one parameter σ is used for the discussion of conductive loss. However, in the Nyquist diagram, σ must be a complex number with real and imaginary parts. If these parts have the same value, the low-frequency part shown in **Figure 9** should be a straight line with a 45° slope, but the measured value is not. Therefore, from this figure, when discussing losses in the microwave band, conductivity must also be considered to have a complex value.

The power density w^* [J/m^3] of one cycle can be calculated from the current density $i(t) = i_0 \sin\omega t$ with amplitude i_0 [A/m] and the electric field intensity $E(t) = E_0 \sin\omega t$ with amplitude E_0 [V/m]. From the definition of the complex conductivity σ^* [S/m] $= \sigma' - j\sigma''$, Eq. (38) is deformed as Eq. (59):

$$i(t) = \sigma^* E(t). \tag{59}$$

Here, w^* is shown as Eq. (60):

$$w^* = \int_0^{2\pi/\omega} i(t)E(t)dt = \frac{1}{2f}\sigma^* E_0^2. \tag{60}$$

Since w^* is repeated f times per second, the energy W received by the conductive material from the electric field per unit volume/unit time is expressed by Eq. (61):

$$W^* = \frac{1}{2}\sigma^* E_0^2. \tag{61}$$

Here, σ^* is a complex value and describes mobility and loss at the same time. When the real part is mobility and the imaginary part is loss, the conduction loss can be described in the same way as Eq. (50), and Eq. (62) is obtained:

$$P_{\sigma\_loss} = \frac{1}{2}\sigma''|E|^2. \tag{62}$$

Comparing Eqs. (50) and (64), it can be seen that σ'' and $\omega\varepsilon_0\varepsilon_r''$ have the same dimensions. Therefore, the loss equation of the combined electric field is Eq. (63):

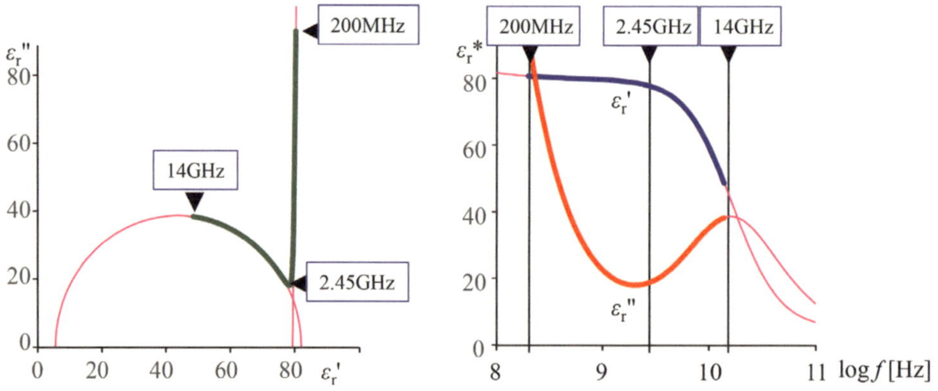

Figure 9.
Nyquist and bode diagrams of 0.1-Mol/L NaCl aqueous solution.

$$P_{E\_loss} = \frac{1}{2}(\sigma'' + \omega\varepsilon'')|E|^2. \tag{63}$$

The apparent relative permittivity measured in **Figure 9** is the sum of the permittivity term and the conductivity term. This response is represented by $\psi_r{}^*(\omega)$. Here, since $\sigma_r{}^*$ has the same dimension as $\omega\varepsilon_r{}^*$, $\psi_r{}^*(\omega)$ is expressed by the following Eq. (64):

$$\psi_r^*(\omega) = \frac{\sigma^*(\omega)}{\omega\varepsilon_0} + \varepsilon_r^*(\omega) = \frac{\sigma_r^*(\omega)}{\omega} + \varepsilon_r^*(\omega). \tag{64}$$

When the complex conductivity σ^* [S/m] = $\sigma' - j\sigma''$ is determined in the same manner as the complex conductivity ε^* [F/m] = $\varepsilon' - j\varepsilon''$, the complex relative permittivity $\varepsilon_r{}^*$ [nd] = $\varepsilon^*/\varepsilon_0$, the complex relative conductivity is derived as $\sigma_r{}^* = \sigma^*/\varepsilon_0 = \sigma_r' - j\sigma_r''$. Here, $\sigma_r{}^*$ has the same dimensions as $\omega\varepsilon_r{}^*$, [rad/s]. Although the international annealed copper standard (IACS) is defined as 58 MS/m as a standard for conductivity, $\sigma_r{}^*$ indicates a ratio to ε_0, rather than IACS.

In the previous diagram of the NaCl aqueous solution, the Nyquist diagram shows a semicircle in the high-frequency band, so this region has dielectric properties. On the other hand, in the low-frequency region, a locus different from a semicircle based on the movement of ions is shown. This means the loss due to the phase delay is small and mainly due to the motion of the ionic molecules. Whether the heat generation behavior at the irradiated frequency is mainly caused by dielectric or conductive loss cannot be distinguished by measurement at one frequency. Unless it is a Nyquist or Bode diagram, the contribution ratio of the dielectric and conductive losses cannot be separated from the locus.

3. Extraordinary microwave effect

In many discussions of microwave chemistry, temperature and heat are very important. It is shown that the loss of electromagnetic energy is based on the imaginary part, i.e. ε_r'' or σ_r'', as shown in Eq. (65). On the other hand, it is stated that energy is accumulated in a material with a large real part because propagation delay occurs, and substantial loss can be obtained by the tan δ term as shown in Eq. (53). Since microwave energy penetrates the material, the amount of energy on the spot can be increased without changing the amount of material. Therefore, it must be remembered that there is a high energy in the field, even if there is no conversion to heat. From this viewpoint, we considered what behavior should be taken if there is a nonthermal effect when microwave energy is applied. This section discusses deductions based on principles, not induction based on data [28, 29].

3.1 Classification of microwave effects

When the chemical reaction field under microwave irradiation is different from the non-irradiation condition, its form can be classified into four types.

1. Fast reaction rate (acceleration);

2. Slow reaction rate (deceleration);

3. Different products (selective production);

4. Different consumptions (selective consuming).

The first is an example in which the reaction speed increases when microwave energy is applied, and as a result, the reaction's end time is shortened. Since the reaction rate can be significantly accelerated by increasing the temperature, one can always discuss whether the temperature was accurately measured or whether the actual reaction field temperature was high. The second is an example opposite to the first. This phenomenon appears when the actual reaction field temperature is low, but this case has little detailed discussion. If phase transition is included in the category, supercooling and overheating phenomena correspond to this. The third is an example in which products differ depending on the presence or absence of microwave irradiation when multiple products are considered. This happens when there are multiple reaction paths, one of which is particularly accelerated. This includes cases where only intermediates are obtained in a multistep reaction. The fourth is an example of a case with multiple substrates, and a reaction of the specific substrates is prioritized. In any of these cases, the reaction rate is considered to have changed due to the application of microwave energy. That is,

1. The target reaction speed increased;

2. The inhibition reaction speed increased;

3. The reaction speed for obtaining target products increased;

4. The reaction speed for consuming specific substrates increased.

All four of these interpretations mean that the specific reaction rate was increased by the application of microwave energy. From the aforementioned, when the chemical reaction under microwave irradiation is different from that in the non-irradiation case, if it is not a thermal effect, an equation for changing the reaction rate must be derived even if the temperature T is constant. The following describes the possibility of such an expression.

3.2 Real and imaginary parts

Eq. (63) describes the attenuation of the energy of the electric field. Considering the propagation here, it is an expression in which the imaginary part is changed to a real part. This means that the propagation equation is derived as Eq. (65):

$$P_{E\_prop} = \frac{1}{2}(\sigma' + \omega\varepsilon')|E|^2. \tag{65}$$

Since the conductive material can be regarded as a special state of the dielectric, discussion of the material including σ will be omitted hereafter. Energy of electromagnetic waves vibrate between electric and magnetic fields as follows equations (Eqs. (66)–(69)). Herein, the magnetic field loss and the propagation energy P_H are also expressed based on the complex permeability $\mu^* = \mu' - j\mu''$ [H/m]:

$$P_{E\_prop} = \frac{1}{2}\omega\varepsilon'|E|^2; \tag{66}$$

$$P_{E\_loss} = \frac{1}{2}\omega\varepsilon''|E|^2; \tag{67}$$

$$P_{H\_prop} = \frac{1}{2}\omega\mu'|H|^2; \tag{68}$$

$$P_{H\_loss} = \frac{1}{2}\omega\mu''|H|^2. \tag{69}$$

Therefore, propagation of electromagnetic waves occurs according to the following stages:

- E-stage 1: Electric field energy given in the system is accumulated as propagation amount P_{E_prop}, and loss amount P_{E_loss} is converted into isotropic thermal motion of the molecules;
- E-stage 2: Propagation amount P_{E_prop} changes to magnetic field energy;
- H-stage 1: The magnetic field energy given in the system is accumulated as propagation amount P_{H_prop}, and the loss amount P_{H_loss} is converted into isotropic thermal motion of the molecules;
- H-stage 2: Propagation amount P_{H_prop} changes to electric field energy.
- These stages are repeated. What this equation indicates is transition of the electromagnetic wave energy, not a chemical reaction; thus, to consider the effect on the chemical reaction, one uses the following quantities:
- P_{X_prop} indicates the amount of energy present in the field acting on the reaction;
- P_{X_loss} indicates the amount of energy lost to the field affecting the reaction;
- A combination of these.

Whether P_{X_prop} is involved as a field that affects the reaction or P_{X_loss} is involved as a result of affecting the reaction is not clearly determined now. In any case, however, it should be energy terms which affect the chemical reaction.

3.3 Interpretation of the Arrhenius equations

In the reaction kinetics, the activation energy theory described by Arrhenius (Eq. (70)), which originated from gas molecular kinetics, is discussed:

$$k = Ae^{-\frac{E_a}{RT}}. \tag{70}$$

Strictly speaking, it cannot be applied theoretically except for the secondary reaction of two gas molecules, but it is useful as an empirical formula and can be used in a solution system. Here, k, A, E_a, and R are the reaction rate constant, A-factor,

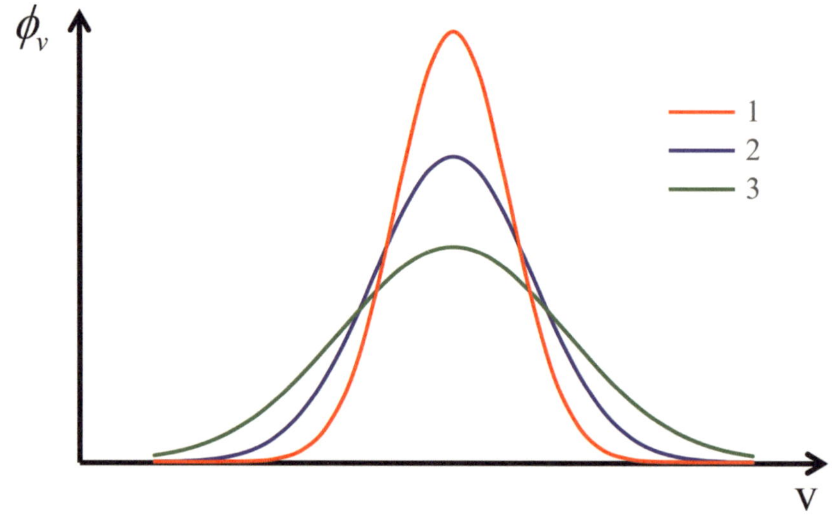

Figure 10.
Relationship between molecular velocity v and existence probability φ_E*. Line 1: small distribution ~ Line 3: large distribution.*

activation energy [J/mol], and molar gas constant [J/Kmol]. E_a is assumed to be a constant that does not change with the reaction temperature.

The Arrhenius equation can be interpreted as follows. The horizontal axis in **Figure 10** shows the molecular velocity v in an arbitrary reaction coordinate system. The vertical axis φ_v represents the existence probability of the molecule with a normal distribution. In **Figure 10**, it is assumed that a molecule that thermally and isotropically exchanges kinetic energy at a temperature T has a normal velocity distribution according to Eq. (71):

$$\phi_v = Ae^{-\frac{mv_x^2}{2RT}}. \tag{71}$$

Here, the mass per mol is m [kg/mol], the kinetic energy of the molecule is $E_v = mv^2/2$ [J/ mol], and the horizontal axis is energy, E. If the range of E is $0 \leq E < +\infty$, this frequency distribution φ_E is canonical, as shown in **Figure 11**.

$$\phi_E = Ae^{-\frac{E}{RT}} \tag{72}$$

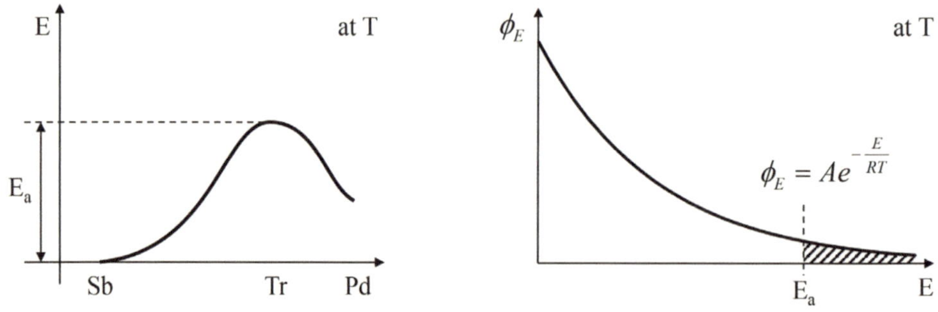

Figure 11.
Reaction coordinate and energy distribution at temperature T.

The progress of the reaction is assumed to occur when the motion of the original molecule coincides with the positive direction of the reaction coordinate system and its kinetic energy exceeds E_a. The reaction coordinate is shown on the left-hand side of **Figure 11**. Here, the substrate, transition state, and product are Sb, Tr, and Pd, respectively. The integrals of the region $E_a < E < \infty$ and of the whole region $0 \leq E < \infty$ are expressed by Eqs. (73) and (74):

$$\int_{E_a}^{\infty} Ae^{-\frac{E}{RT}}dE = ARTe^{-\frac{E_a}{RT}}; \tag{73}$$

$$\int_{0}^{\infty} Ae^{-\frac{E}{RT}}dE = ART. \tag{74}$$

Therefore, the ratio of the region exceeding E_a to the entire region is represented by Eq. (75), indicating that A is irrelevant:

$$\frac{\int_{E_a}^{\infty} Ae^{-\frac{E}{RT}}dE}{\int_{0}^{\infty} Ae^{-\frac{E}{RT}}dE} = e^{-\frac{E_a}{RT}}. \tag{75}$$

The value obtained by Eq. (75) is the Boltzmann factor. In the Arrhenius equation, the reaction rate is proportional to the Boltzmann factor because of the occupation ratio of the high energy state.

Here, the temperature T of the heat bath is increased to $T + \Delta T$. In gas molecule kinetics, the kinetic energy of a molecule increases with temperature, and this is expressed by $(m_B v^2)/2 = 3k_B T/2$. Here, m_B and k_B are the mass of one molecule and the Boltzmann constant, respectively. Therefore, a temperature increase of ΔT is the same as one molecule receiving $k_B \Delta T/2$ of energy as $m_B v_x^2/2$, $m_B v_y^2/2$, and $m_B v_z^2/2$, respectively. Considering this at 1 mol, since $R = N_A k_B$ (where N_A is Avogadro's number), the kinetic energy of the system is distributed to the x, y, and z components by $R\Delta T/2$, respectively.

When any one direction is taken as the reaction coordinate, the energy distribution is the same in any state from Sb to Pd in the reaction coordinate system. Therefore, as expected, E_a does not change, even if the temperature is raised. This is shown on the left-hand side of **Figure 12**. On the other hand, an increase in T corresponds to a decrease in the kurtosis of the Boltzmann distribution in the original molecule and an increase in the tail. As a result, the occupancy rate of molecules having an energy exceeding E_a increases, and the reaction rate k increases, as shown at right in **Figure 12**.

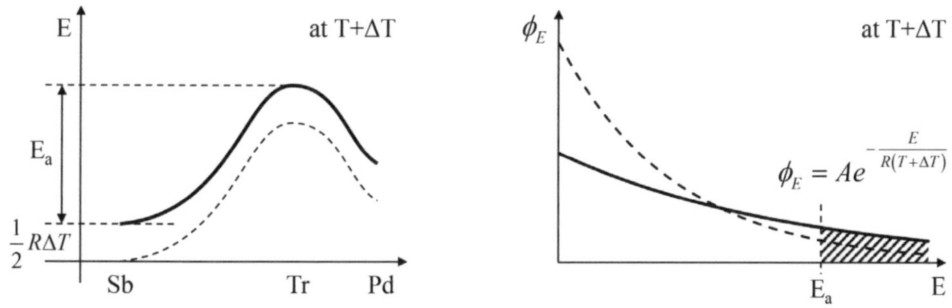

Figure 12.
Reaction coordinate and energy distribution at temperature T + ΔT.

Next, consider the supply of microwave energy instead of ΔT as an external energy supply. The Arrhenius equation presupposes that it is in thermal equilibrium, which indicates that energy can be exchanged quickly with an external heat bath. Therefore, if the rate at which the system releases heat to the outside matches the heating rate by microwaves, a constant temperature state can be maintained, and T can be regarded as constant. This means that there can be reaction systems with the same T and different microwave application intensities. Microwave energy is calculated as energy per unit time. On the other hand, the reaction kinetics discussed earlier calculate the change in the molar concentration of molecules as the reaction rate. Therefore, in order to align the units, the electromagnetic wave energy supplied to the volume per mole with respect to the concentration M [mol/m^3] of the target reaction molecule is E_{add} [J/mol]. E_{add} is proportional to the oscillated energy E_{MW} [J/mol], which is the product of power per mole and irradiation time, but not all energy works effectively.

Herein, the effective efficiency is defined by α and γ, and $E_{add} = \alpha RT = \gamma E_{MW}$. This means that αRT is added to the thermal potential RT that the field already has. Assuming that this added amount of external energy is incorporated into the canonical distribution in the same manner as RΔT, the potential distribution is derived in Eq. (76) by replacing RT in Eq. (72) with $(1 + \alpha)RT$:

$$\phi_E = A_1 e^{-\frac{E}{(1+\alpha)RT}} = A_1 e^{-\frac{E}{RT+E_{add}}}. \tag{76}$$

The coefficient A_1 in Eq. (76) was introduced to consider the possibility that the A-factor changes depending on the presence or absence of external energy.

From the right-hand side of **Figure 13**, integration of the region $E_a < E < \infty$, integration of the whole region $0 \le E < \infty$, and the ratio between both are obtained as Eqs. (77), (78), and (79):

$$\int_{E_a}^{\infty} A_1 e^{-\frac{E}{(1+\alpha)RT}} dE = (1+\alpha)A_1 RT e^{-\frac{E_a}{(1+\alpha)RT}}; \tag{77}$$

$$\int_{0}^{\infty} A_1 e^{-\frac{E}{(1+\alpha)RT}} dE = (1+\alpha)A_1 RT; \tag{78}$$

$$\frac{\int_{E_a}^{\infty} A_1 e^{-\frac{E}{(1+\alpha)RT}} dE}{\int_{0}^{\infty} A_1 e^{-\frac{E}{(1+\alpha)RT}} dE} = e^{-\frac{E}{(1+\alpha)RT}} = e^{-\frac{E}{RT+E_{add}}}. \tag{79}$$

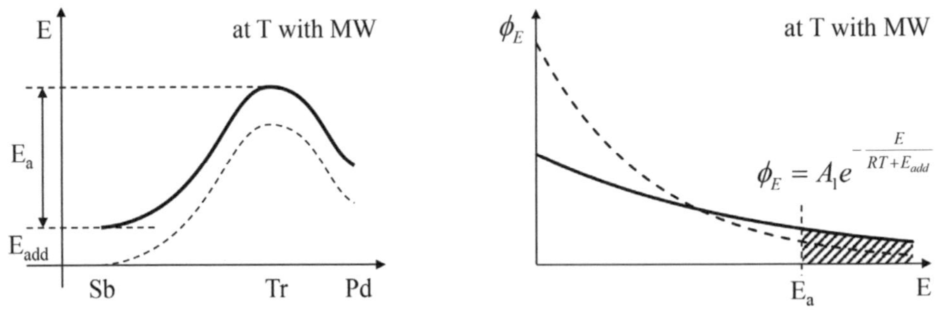

Figure 13.
Reaction coordinate and energy distribution at temperature T *under microwave irradiation.*

From the aforementioned, it is apparent that the exponential form of the Boltzmann factor is maintained, and this expression is irrelevant to A_1, just as was Eq. (75). Here, when the reaction rate constant under microwave non-irradiation is k_0 and that under irradiation is k_1, the Arrhenius equation is described as follows:

$$k_0 = A_0 e^{-\frac{E_a}{RT}}; \tag{80}$$

$$k_1 = A_1 e^{-\frac{E_a}{RT + E_{add}}}. \tag{81}$$

Comparing both equations, $k_0 = k_1$ at $E_{add} = 0$. Therefore, $A_0 = A_1$ is derived. Eq. (81) shows that the A-factor does not change under microwave irradiation.

3.4 Introduction of a nonthermal constant

Generalizing Eq. (81) under an applied microwave irradiation yields Eq. (82):

$$k = Ae^{-\frac{E_a}{RT + \gamma E_{MW}}}. \tag{82}$$

E_{MW} is an intensive property that can change the applied amount from the outside. Therefore, changing k by a variable independent of temperature T is not thermal, that is, a nonthermal effect. Therefore, the coefficient γ in Eq. (82) is a nonthermal constant. Here, if γ is a positive value, the group has the same T, but the variance of the canonical distribution widens, and the occupation ratio of E_a or higher is increased and $k_0 < k_1$.

Eq. (82) does not change E_a and A. Therefore, the reaction route is not changed. This is a reaction rate equation that can be applied, even when the microwave irradiation is 0, and the irradiation intensity of the microwave is variable. In this case, the population maintains a canonical distribution. If there is a nonthermal effect, it must be expressed by a reaction equation with a microwave intensity independent of temperature as a variable. Eq. (82) obtained by deduction meets this condition.

3.5 Working principle of a nonthermal constant

Eqs. (81) or (82) can be transformed into Eq. (83). E_a and A can be obtained without microwave irradiation. If E_a and A are not changed by microwave irradiation, E_{add}, that is, γE_{MW} can be obtained from k at T. Assuming that the volume of the irradiation target is same and E_{MW} is proportional to the irradiation power, γ can be calculated from Eq. (83):

$$\alpha RT = E_{add} = \gamma E_{MW} = \frac{E_a}{\ln(A) - \ln(k)} - RT. \tag{83}$$

If the effect of microwave irradiation is expressed by Eq. (81), the ratio r between Eqs. (81) and (80) indicates the efficiency of increase by microwave irradiation (Eq. (84)):

$$r = \frac{Ae^{-\frac{E_a}{(1+\alpha)RT}}}{Ae^{-\frac{E_a}{RT}}} = e^{\left(\frac{\alpha}{1+\alpha}\right)\frac{E_a}{RT}}. \tag{84}$$

When α is 0, there is no irradiation, so $r = 1$. The upper limit of r is considered when α is infinite. Therefore, the upper limit $r_{\max}$ of r is expressed by Eq. (85):

$$r_{max} = \lim_{a \to \infty} e^{\left(\frac{\alpha}{1+\alpha}\right)\frac{E_a}{RT}} = e^{\frac{E_a}{RT}}. \tag{85}$$

This equation suggests that in a system having an effective α, the effect of microwave irradiation appears more markedly as the reaction temperature T is lower and E_a is higher.

4. Conclusions

Microwave chemistry was described with a focus on the behavior of dielectrics. We discussed the relationship between physical properties such as dielectric constant, energy propagation in the material, and energy loss in the material. By discussing energy, if there is a special effect other than heat in microwave chemical reactions, we have derived *a priori* by mathematical formulas factors that are required. The "Arrhenius equation under microwave irradiation" proposed in this section is one model. In this equation, a microwave energy term is added.

In order to confirm the results of this deduction inductively, correct measurements are required. A simple method is a surface temperature measurement with a radiation thermometer, but the fact that there is a temperature difference from the inside has been investigated in various experiments, and unless this problem is solved, it will lead to incorrect measurement. The temperature and its distribution should be validated with multiple methods, such as internal radiation measurements using an optical fiber, a fiber-grating method that measures the local volume changes of the optical fiber, and temperature dependence of the lifetime of the fluorescent material at the tip [30–32].

As presented in this paper, understanding microwave effects algebraically in terms of energy theory is different from considering energy distribution geometrically. For example, the existence of nonequilibrium local heating, which is influenced by the structure of the irradiated object, has been reported [33]. This phenomenon, which can be explained as a peculiar heating method occurring in microwave heating, is due to the geometric intensity of the electromagnetic field distribution. In this example, the unique heat distribution structure gives "geometric" microwave effects rather than "algebraic" microwave effects.

Microwaves in the GHz band have a decimator wavelength. If the size of the reaction vessel is smaller than the wavelength, the result may vary greatly depending on the place of installation and the shape of the vessel, based on the microwave interference. Therefore, in chemical synthesis assisted by microwave heating, it is necessary to consider the shape of the electromagnetic field distribution in the apparatus. Otherwise, other conditions may change at the same time that the temperature, substrate, solvent, and scale are changed. If this unintended influence is ignored, it becomes difficult to clarify general trends in the condition search, or the result will lead to excessively good or bad evaluations. Thus, microwave chemistry seems to be complicated in terms of chemistry and electromagnetic field analysis. However, it can be expected that more efficient reaction control will be possible by fully utilizing the control as an external field and utilizing it highly. In addition to the algebraic interpretation based on energetics, if the structure of the irradiated object is geometrically controlled as a metamaterial, further microwave effects will be manifested. We believe that microwave chemistry will be a useful technique that can be used to manipulate chemical synthesis by applying external energy to a simple heating reaction.

Author details

Jun-ichi Sugiyama*, Hayato Sugiyama, Chika Sato and Maki Morizumi
National Institute of Advanced Industrial Science and Technology (AIST), Tsukuba, Ibaraki, Japan

*Address all correspondence to: sugiyama-j@aist.go.jp

References

[1] Ritter SK. Microwave chemistry remains hot, fast, and a tad mystical. Chemical & Engineering News. 2014; **92**(4):26-28

[2] Kappe A, Pieber B, Dallinger D. Effects in organic synthesis: Myth or reality? Angewandte Chemical International Edition. 2013;**52**: 1088-1094. DOI: 10.1002/anie.201204103

[3] Rosana MR, Tao Y, Stiegman AE, Dudley GB. On the rational design of microwave-actuated organic reactions. Chemical Science. 2012;**3**:1240-1244. DOI: 10.1039/c2sc01003h

[4] Dudley GB, Stiegman AE, Rosana MR. Correspondence on microwave effects in organic synthesis. Angewandte Chemie, International Edition. 2013;**52**:7918-7923. DOI: 10.1002/anie.201301539

[5] Kappe CO. Reply to the correspondence on microwave effects in organic synthesis. Angewandte Chemie, International Edition. 2013;**52**: 7924-7928. DOI: 10.1002/anie.201304368

[6] Nushiro K, Kikuchi S, Yamada T. Microwave effect on catalytic enantioselective Claisen rearrangement. Chemical Communications. 2013;**49**: 8371-8373. DOI: 10.1039/c3cc44610g

[7] Sugiyama J. Heating principle of microwave from a viewpoint of the mathematical formula. In: Proceedings of the 4th JEMEA Symposium. Fukuoka. Tokyo: JEMEA; 2010. pp. 26-27

[8] The International System of Units. 9th ed. The International Bureau of Weights and Measures; 2019 ISBN 978-92-822-2272-0

[9] Sugiyama J. Change of complex permittivity and electromagnetic field accompanying with temperature rising. In: Proceedings of the JEMEA Safety/technology Seminar. Fukuoka. Tokyo: JEMEA; 2010. pp. 40-50

[10] Sugiyama J, Morizumi M, Sato C. Separation of conductivity sigma and permittivity epsilon in electrolyte solution. IEICE Technical Report. 2014; **MW2014**:1-6

[11] Sugiyama J. Development of the resonator to measure a complex permittivity at the wide temperature. IEICE Technical Report. 2009; **MW2009-79**:31-36

[12] Sugiyama J. Evaluation of the phase displacement delta by the complex permittivity measuring with resonators. IEICE Technical Report. 2010;**MW2009**: 11-16

[13] Sugiyama J. Calculation of the relaxation time tau by resonators and thermal behavior in heating device. IEICE Technical Report. 2010;**MW2010**: 13-18

[14] Sugiyama J. What is the microwave heating from a viewpoint of a material? JEMEA-Bulletin. 2017;**3**(1):15-18

[15] Sugiyama J. Heating principle of microwave from a viewpoint of the mathematical formula. In: Proceedings of the 7th JEMEA Symposium. Fukuoka. Tokyo: JEMEA; 2013. pp. 148-149

[16] Sugiyama J, Morizumi M, Sato C. Frequency dependence of the dielectric-relaxation of single liquid. In: Proceedings of the 7th JEMEA Symposium. Tokyo: JEMEA; 2013. pp. 188-189

[17] Sugiyama J. Microwave heating from the viewpoint of materials. In: Proceedings of the 1st JEMEA Summer School. Takayama. Tokyo: JEMEA; 2016. pp. 35-80

[18] Sugiyama J. Microwave Course I, II, III. In: Proceedings of the 13th JEMEA Symposium. Tsukuba. Tokyo: JEMEA; 2019. p. 28

[19] Gabriel C, Gabriel S, Grant EH, Halstead BSJ, Mingos DMP. Dielectric parameters relevant to microwave dielectric heating. Chemical Society Reviews. 1998;**27**(3):213-223. DOI: 10.1039/a827213z

[20] Sugiyama J, Morizumi M, Sato C. Frequency dependence of the dielectric-relaxation of blended liquid. In: Proceedings of the 7th JEMEA Symposium. Tokyo: JEMEA; 2013. pp. 190-191

[21] Sugiyama J, Sato C. Heating of small amount solution by a solid oscillator. In: Proceedings of the 10th JEMEA Symposium. Tokyo: JEMEA; 2016. pp. 160-161

[22] Sugiyama H, Sugiyama J, Sato C. Study of characteristic correlation for dielectric liquid mixture. In: Proceedings of the 13th JEMEA Symposium. Tokyo: JEMEA; 2019. pp. 162-163

[23] Sugiyama J, Morizumi M, Sato C. Mathematical analysis of the empirical relationship of the dielectric-relaxation. In: Proceedings of the 8th JEMEA Symposium. Kochi, Tokyo: JEMEA; 2014. pp. 74-75

[24] Miyamoto S, Sugiyama J. Microwave characteristics measurement of the liquid which has electrical conductivity sigma. IEICE Technical Report. 2011; **MW2011**:13-18

[25] Sugiyama J, Yamazaki T, Moriike T, Suzuki M, Segawa T, Kato Y, et al. Prototyping of large-scale circular microwave oven and its actual heating efficiency at complete evaporation of water. IEICE Technical Report. 2011; **MW2010**:51-56

[26] Nagashima I, Sugiyama J, Sakuta T, Sasaki M, Shimizu H. Efficiency of 2.45 and 5.80 GHz microwave irradiation for a hydrolysis reaction by thermostable β-glucosidase HT1. Bioscience, Biotechnology, and Biochemistry. 2014; **78**(5):758-760. DOI: 10.1080/09168451.2014.891931

[27] Sugiyama J, Morizumi M, Sato C. Evaluation of the real and imaginary part of complex conductivity. In: Proceedings of the 8th JEMEA Symposium. Kochi, Tokyo: JEMEA; 2014. pp. 198-199

[28] Sugiyama J. Proposition of theory on the quantitativity of non-thermal effects. In: Proceedings of the 13th JEMEA Symposium. Tsukuba, Tokyo: JEMEA; 2019. pp. 41-42

[29] Sugiyama J, Yoneya M. Simulation of molecular motion in the application of alternating electromagnetic fields in the microwave band. Computation of Chemical Japan. 2021;**20**(3):123-125. DOI: 10.2477/jccj.2021-0035

[30] Sugiyama J. Electromagnetic relationship between microwaves and flow reactor systems. Chemical Record. 2019;**19**:146-156. DOI: 10.1002/tcr.201800120

[31] Wada D, Sugiyama J, Zushi H, Murayama H. An optical fiber sensing technique for temperature distribution measurements in microwave heating. Measurement Science and Technology. 2015;**26**(085105):1-7. DOI: 10.1088/0957-0233/26/8/085105

[32] Wada D, Sugiyama J, Zushi H, Murayama H. Temperature distribution monitoring of a coiled flow channel in microwave heating using an optical fiber sensing technique. Sensors and Actuators B: Chemical. 2016;**232**: 434-441. DOI: 10.1016/j.snb.2016.03.156

[33] Wada Y, Tsubaki S, Maitani MM, Fujii S, Kishimoto F, Haneishi N. Physical insight to microwave special effects: Nonequilibrium local heating and acceleration of Electron transfer. Journal of the Japan Petroleum Institute. 2018;**61**(2):98-105. DOI: 10.1627/jpi.61.98

Chapter 5

Power Consumption in CMOS Circuits

Len Luet Ng, Kim Ho Yeap, Magdalene Wan Ching Goh and Veerendra Dakulagi

Abstract

In this chapter, we explain the two types of power consumption found in a complementary metal-oxide-semiconductor (CMOS) circuit. In general, a CMOS circuit tends to dissipate power at all times—be it active or inactive. The power consumed by the circuit when it is performing computational tasks is known as dynamic power. On the contrary, the power lost due to current leakage during which the circuit is dormant is referred to as static power. By carefully and properly designing the circuit, current leakage can be suppressed to its minimum. Hence, dynamic power consumption is usually significantly higher than its static counterpart. Some of the techniques that could be adopted to save dynamic power consumption include reducing the supply voltage, clock frequency, clock power, and dynamic effective capacitance. By probing into the activity factors of the design modules, the techniques can be applied to those with high power consumption.

Keywords: dynamic power, static power, switching power, short-circuit power, leakage power, supply voltage, clock frequency, dynamic effective capacitance, switching activity

1. Introduction

More than half a century has elapsed since the three physicists from the AT&T Bell Laboratories—Brattain, Bardeen, and Shockley—invented the first solid-state transistor in December 1947 [1–3]. In comparison with the thermionic triode (which is colloquially known as the vacuum tube), the solid-state transistor is much smaller in size, consumes much lower power, operates at a relatively lower temperature, and exhibits significantly faster response time. Hence, the solid-state transistor swiftly replaced its predecessor as the predominant building block for electronic devices. The inexorable widespread application of solid-state transistors in electronic circuits has triggered a dramatic revolution in the electronic industries.

Today, microchips are built from the solid-state metal-oxide-semiconductor field-effect transistors (MOSFETs). A typical microchip consists of arrays of negative and positive MOSFETs, which are commonly denoted as the NMOS and PMOS transistors, respectively. **Figures 1** and **2** illustrate the symbols and cross-sections of the NMOS and PMOS transistors. As can be seen from the figures, the source and drain terminals

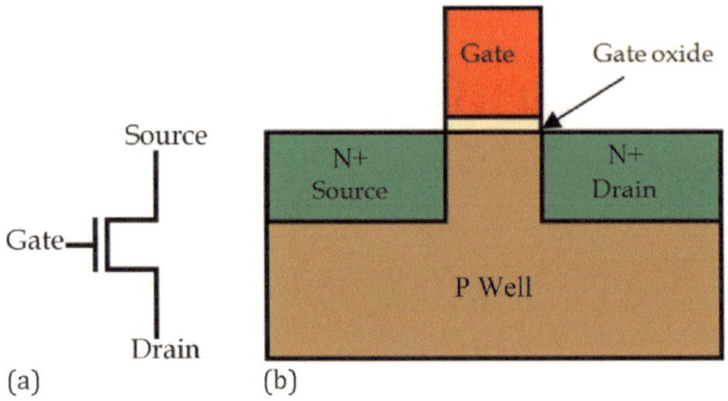

Figure 1.
The (a) symbol and (b) cross-section of a NMOS transistor.

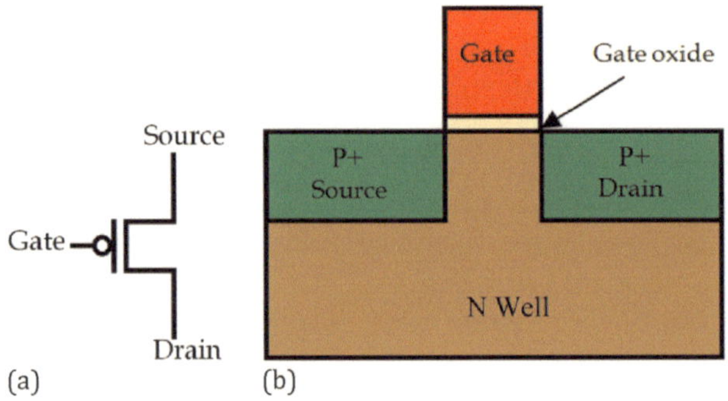

Figure 2.
The (a) symbol and (b) cross-section of a PMOS transistor.

of the NMOS transistor are heavily doped with donator ions, such as phosphorous (P) or arsenic (As), whereas, its body is moderately doped with boron (B) acceptor ions. The PMOS transistor, on the other hand, consists of a high density of B ions at its source and drain and a moderate density of P ions at its body. Since the combination of these two transistors dissipates lower static power and offers higher noise immunity than implementing either the NMOS or PMOS transistor alone, they are both applied concurrently when designing electronic circuits. An electronic circuit that constitutes both the NMOS and PMOS transistors is referred to as the complementary metal-oxide-semiconductor or CMOS.

The insatiable desire to incorporate more functionalities into a microchip has issued a clarion call for a higher number of transistors to be fabricated within it. A state-of-the-art electronic device today, for instance, may be equipped with the fifth-generation (5G) telecommunication, neural engine (NE), augmented reality (AR), cloud computing, facial and speech recognition, and wireless power transmission technologies. These features could only be supported by millions, if not billions, of transistors in the chip. In order to build more transistors into the chip, the size of a transistor has undergone significant reductions over the years [4–6]. By shrinking the size of the transistor, the switching speed of the logic components can also be enhanced, while the operating power can be saved [7]. An advanced microprocessor today possesses more than 50 billion transistors, with technology nodes as small as 5 nm, clock rates of about 5 GHz [3], and an area less than 500 mm^2. Microchips are now interwoven seamlessly with the fabric of mankind, and, in many aspects, they have become an indispensable necessity to mankind.

2. Power consumption in a CMOS circuit

The total power consumption P_{total} in a CMOS circuit comprises two major components, namely, the dynamic power $P_{dynamic}$ and static power P_{static}, that is,

$$P_{total} = P_{dynamic} + P_{static} \tag{1}$$

$P_{dynamic}$ refers to the power consumed by the circuit when it is performing useful work during the active mode, whereas P_{static} is the power lost due to the leakage current that flows through the transistors when the circuit is inactive [8]. An overview of the different types of power consumption is displayed in **Figure 3**.

2.1 Dynamic power consumption

A CMOS circuit dissipates dynamic power $P_{dynamic}$ in either of the following conditions:

i. When there are switching activities at the nodes. The switching activity (SA) refers to the change of the logic state and the probability that the circuit node switches its state from logic 0 to logic 1 and vice versa, which is known as the activity factor (AF). Clearly, the circuit consumes a higher switching power P_{switch} when the frequency of the transistors toggle increases (i.e., the transistors consist of higher AF).

ii. When both NMOS and PMOS transistors conduct current during signal transitions. When the logic changes its state, there is a short period of time when the PMOS and NMOS transistors are switched on simultaneously. The circuit is, therefore, temporarily short-circuited, resulting in power dissipation P_{short}.

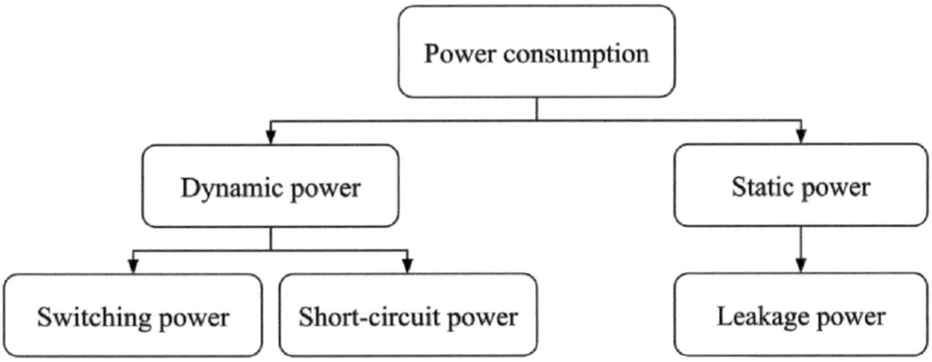

Figure 3.
Different types of power consumption in a CMOS circuit.

Dynamic power dissipation due to the transient short-circuit current path is considerably lower than that caused by the circuits with high switching activities. Hence, emphasis is usually given on finding ways to reduce P_{switch}. Nevertheless, the noise created by the short-circuit current may sometimes be disturbing since it could cause errors in the output logic.

2.1.1 Switching power

In general, the energy delivered to a CMOS circuit can be classified into two parts, namely, the charging and discharging of the load capacitance C_L. To understand how energy delivery takes place, a simple CMOS inverter is shown in **Figure 4**, which is used for illustration.

During the charging phase, the input gate signal switches from logic 1 to 0, and, as a result, the PMOS transistor is switched on, while its NMOS counterpart is switched off. As can be seen in **Figure 4(a)**, the load capacitance C_L is connected to the supply voltage via the PMOS transistor, thereby allowing current $I(t)$ to charge C_L to the supply voltage V_{DD}. The energy E_d delivered to C_L is derived in Eq. (2) given below:

$$E_d = \int_0^\infty I(t) V_{DD}\, dt. \tag{2}$$

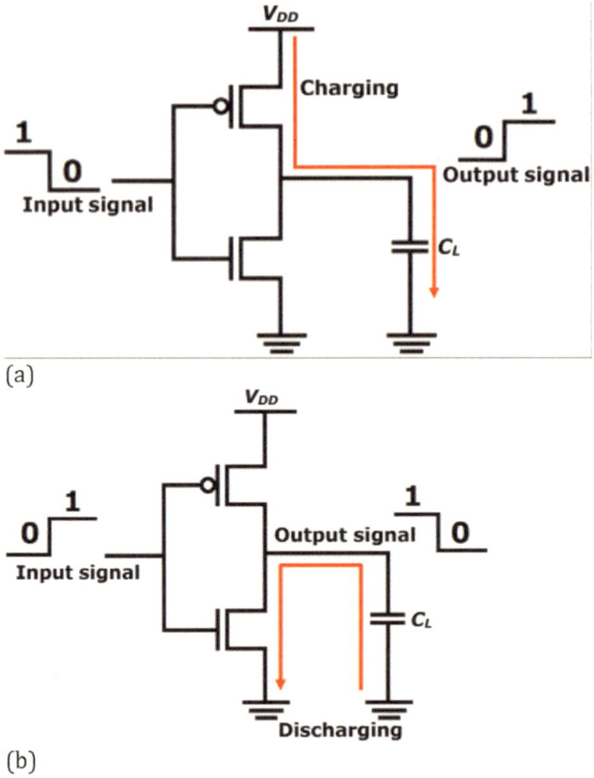

Figure 4.
The (a) charging path (V_{DD} to C_L) and (b) discharging path (C_L to GND) of the capacitive load in the CMOS circuit.

Since the current to charge a capacitor C to voltage V can be obtained from

$$I(t) = C\frac{dV}{dt}, \tag{3}$$

E_d in Eq. (2) can, therefore, be expressed as [8].

$$E_d = C_L V_{DD}^2. \tag{4}$$

Likewise, the energy stored E_c in C_L during the charging phase for each transition is derived in Eq. (5) given below:

$$E_c = \int_0^{\infty} I(t)V(t)dt = \frac{1}{2}C_L V_{DD}^2 \tag{5}$$

It can be observed between Eqs. (4) and (5) that only half of the delivered energy E_d is stored in C_L, while the remaining half is dissipated in the PMOS transistor. In other words, a CMOS circuit encounters power loss for each logic transition when the current passes the transistors. The changing of the logic state is known as the switching activity. Each time a switching activity occurs for a particular node, the transistors will consume energy. Hence, Eq. (4) rather than Eq. (5) is to be used when determining the switching power consumption of a CMOS circuit. This is because both the energy stored in the load and that dissipated in the transistors have to be taken into account.

When the gate signal changes from logic 0 back to 1, the opposite scenario as that of the charging phase occurs this time, the PMOS transistor is switched off and NMOS switched on. During this discharging phase, the energy stored previously in the load capacitance $E_c = \frac{1}{2}C_L V_{DD}^2$ is drained completely to the ground via the NMOS transistor, as seen in **Figure 4(b)**.

When deriving Eq. (4), only a single state transition is considered. In reality, however, the scenario of the signal change at the circuit node may be more complicated than that. Since the feeding signal may have more than one transition within a time interval, Eq. (4) has to be multiplied by N times of transitions to obtain the total delivered power E_{dt}, that is,

$$E_{dt} = N \cdot C_L V_{DD}^2. \tag{6}$$

Assuming that a circuit node toggles at frequency f_{switch} over a time interval T, N can be written as

$$N = T \cdot f_{switch}. \tag{7}$$

Substituting Eq. (7) into (6), the total energy delivered can be expressed as

$$E_{dt} = T \cdot f_{switch} C_L V_{DD}^2. \tag{8}$$

Since power is defined as the rate at which energy is used, the relationship between the switching power P_{switch} and the total delivered power E_{dt} can be described as

$$P_{switch} = \frac{E_{dt}}{T}. \tag{9}$$

Substituting Eq. (8) into (9), the switching power P_{switch} can be written as

$$P_{switch} = f_{switch} C_L V_{DD}^2. \quad (10)$$

Eq. (10) can only be used to accurately calculate P_{switch} as long as the assumption in Eq. (7) holds valid. In most CMOS circuits, however, logic does not really switch at a constant frequency f_{switch}. It is, therefore, more persuasive to express f_{switch} in terms of the product of AF and the clock frequency f_{clk}, that is,

$$f_{switch} = AF \cdot f_{clk} \quad (11)$$

Doing so, P_{switch} in Eq. (10) becomes

$$P_{switch} = AF \cdot f_{clk} \cdot C_L V_{DD}^2 \quad (12)$$

The product of $C_L \cdot AF$ is typically represented by the variable C_{dyn}, which is called the dynamic effective capacitance. Hence, Eq. (12) can be rewritten as

$$P_{switch} = f_{clk} \cdot C_{dyn} V_{DD}^2. \quad (13)$$

According to Weste and Harris [9], activity factor (AF) is defined as the probability that the circuit node changes from logic 0 to logic 1, and this is the only time that the circuit consumes switching power. Therefore, AF is an important element to estimate the power consumption of a circuit. In order to gain a better understanding of AF, assume that the clock is triggered at every single cycle, that is, $f_{switch} = f_{clk}$. From Eq. (11), it can be seen that $AF = 1$. Likewise, AF is found to be 2 for a data signal, which toggles once every two clock cycles. This phenomenon is graphically depicted in **Figure 5**. In most cases, the least significant bit (LSB) and the most significant bit (MSB) contain the highest and lowest AF, respectively.

For a circuit node that switches its logic states in an irregular manner, AF can be determined by multiplying the probability the node switches to logic 0, P_f^0, with the probability it switches to logic 1, P_f^1, that is,

$$AF = P_f^0 \times P_f^1 \quad (14)$$

For instance, a switching function expression is given as $F = \overline{A + BC + \overline{B}C}$. The truth table of F is shown in **Table 1**. Out of the eight combinations of input values in

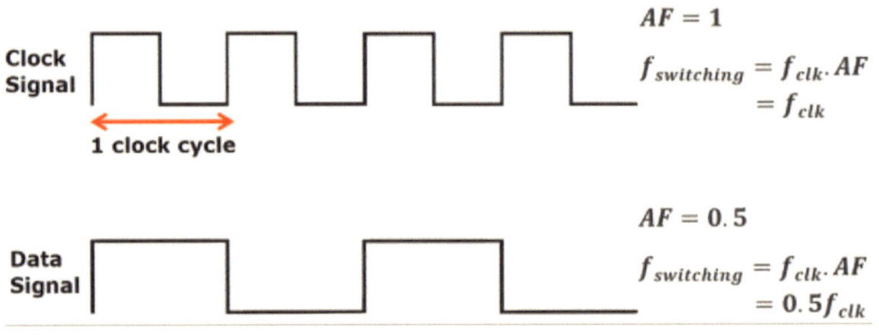

Figure 5.
Clock signal (AF = 1) *and data signal* (AF = 0.5).

A	B	C	F(A, B, C)
0	0	0	0
0	0	1	1
0	1	0	1
0	1	1	0
1	0	0	0
1	0	1	0
1	1	0	0
1	1	1	0

Table 1.
Truth table of $F = \overline{A + BC + \overline{B}\,\overline{C}}$.

the truth table, two produce logic 1 at output *F*, while the remaining ones produce logic 0. Hence, P_f^0 and P_f^1 can be obtained as $\frac{6}{8}$ and $\frac{2}{8}$, respectively. Substituting these values into Eq. (14), *AF* is then found to be 0.1875. This is to say that, the probability that the circuit is active is only 0.1875, which is clearly low. Comparing this value with the frequency of occurrence for logic 1 found in the truth table, it can be seen that *AF* gives a good indication of the active rate of the circuit.

2.1.2 Short-circuit power

Since it takes time for the parasitic capacitance to charge and discharge, the signal fed to a circuit does not change its logic state instantly. The input signal consists of the finite rise and fall times by this means. Unlike the switching power dissipation, where only one of the transistors is switched on at a time, short-circuit power dissipation P_{short} is induced from the concurrent activation of both the NMOS and PMOS transistors. When the logic changes its state, there is a short window of time where the PMOS and NMOS transistors are switched on simultaneously. A direct current path connecting V_{DD} to the ground is produced within this interval, resulting in short-circuit power dissipation P_{short}. As can be seen in **Figure 6**, the short-circuit current that passes both the PMOS and NMOS transistors during the transition state does not contribute to the charging and discharging of the load capacitance. The power

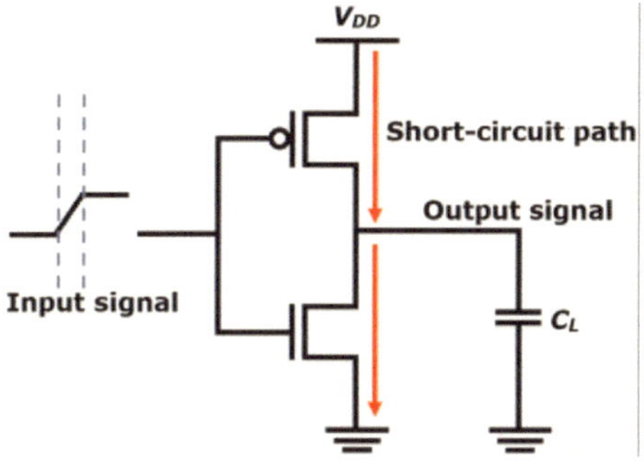

Figure 6.
Short-circuit path (V_{DD} to GND) in a CMOS circuit.

produced does not deliver any meaningful activities at the output and is therefore wasted. The short-circuit power P_{short} can be mathematically expressed as

$$P_{short} = T_{sc} \cdot V_{DD} \cdot I_{peak} \tag{15}$$

where T_{sc} is the rising or falling time of the input signal and I_{peak} is the peak current, which could be estimated from the transistor size and technology process.

2.2 Static power

Static power P_{static} refers to the power lost when the CMOS circuit is dormant. The main culprit of P_{static} is the leakage current, which exists mainly because of the short-channel effects [10]. As the technology node continues to reduce toward the sub-nanometer range, leakage current has become a major problem. In 2011, the severity of the leakage reached the brink, which prompted Intel Corporation to introduce the 22 nm tri-gate transistor. The tri-gate transistor is more popularly referred to as the FinFET, owing to its protruding drain and source structures, which resemble the fin of a fish. In comparison with the planar MOSFETs, the FinFET has better control of the current flow, thereby reducing leakage [11].

Among the short-channel effects that contribute predominantly to the P_{static} dissipation are the sub-threshold leakage current and the gate leakage current [12]. Sub-threshold leakage current is the weak current that exists between the source and drain terminals during the off-state of the transistor, as a result of the weak inversion layer at the oxide–substrate interface. This phenomenon occurs when the gate voltage is lower than the threshold voltage V_{TH}. The sub-threshold leakage increases exponentially as the feature size continues to shrink [13]. This is mainly caused by the reduction of the V_{TH}, which is also scaled down accordingly.

When the size of the transistor decreases, the oxide layer is thinned as well. The oxide thickness is continuously thinned until a certain extent, an undesired electric field is induced at the oxide–substrate interface whenever voltage is applied at the gate terminal. The electric field increases the probability of electrons to tunnel through the oxide layer from the channel region into the gate and vice versa [14], leading to gate leakage current. Although its impact is less severe compared to the sub-threshold leakage, current leakage by virtue of this mechanism has gradually exacerbated when the technology node penetrates the nanometric regime.

The equation that describes static power dissipation can be expressed as,

$$P_{static} = V_{DD} \cdot I_{leakage} \tag{16}$$

where $I_{leakage}$ denotes the total leakage current.

3. Dynamic power optimization

With an increasing number of features being installed into a microchip today, higher computing power is drawn by electronic devices. Power consumption has, therefore, become a key concern in a CMOS circuit designs. Clearly, it is imperative to employ effective approaches to cut down the usage of power in microchips.

The first step for power optimization is to analyze the overall power consumed by a CMOS circuit. By substituting Eqs. (13), (15), and (16) into Eq. (1), the total power consumption P_{total} can be obtained as

$$P_{total} = \left(f_{clk} \cdot C_{dyn} \cdot V_{DD}^{2}\right) + \left(T_{sc} \cdot V_{DD} \cdot I_{peak}\right) + \left(V_{DD} \cdot I_{leakage}\right). \tag{17}$$

Upon inspection, it can be observed that P_{total} is dictated by six power components, namely, f_{clk}, C_{dyn}, V_{DD}, T_{sc}, I_{peak}, and $I_{leakage}$. By carefully and properly designing the logic circuit, P_{short} and P_{static} can usually be minimized, if not completely suppressed. Hence, only clock frequency f_{clk}, dynamic effective capacitance C_{dyn}, and supply voltage V_{DD} are typically adjusted to optimize the power consumption of the circuit. Although modifying any of these three components may result in power saving, precaution is to be heeded since some approaches may also impose adverse effects on the circuits. The reduction of V_{DD}, for example, yields a quadratic effect on diminishing the usage of power. Doing so, however, may also impair the performance of the circuit. This approach is generally not recommended unless there is a need to switch the device from high performing to high power saving. Similarly, reducing the clock frequency f_{clk} may not necessarily be effective in curtailing power consumption because the system has to operate for a longer period of time when fed with the same load. Tuning C_{dyn} is perhaps one of the most popular options since it involves decreasing either the parasitic capacitance or the activity factor AF for idle nodes. A summary of some of the existing methods adopted to optimize power usage is presented in the subsequent sections. The first three methods discussed in the subsequent sections are related to the adjustments of the clock frequency f_{clk} and supply voltage V_{DD}, while the remaining ones deal with the dynamic effective capacitance C_{dyn}.

3.1 Dynamic frequency scaling

Weissel and Bellosa [15] proposed an event-driven clock scaling approach for dynamic power management. In their work, schedulers were utilized to determine the appropriate clock frequency for each thread. Finally, the frequency of the dedicated applications can be adjusted based on the recurrent analysis of the thread-specific performance profile. Their analysis showed that at least 37% of energy can be saved with a performance loss of less than 10% when tested on the Intel XScale architecture.

Although this approach shows positive results in power saving, it may suffer from longer application execution time, since the frequency is reduced. As a consequence of this, applications may be corrupted if certain deadlines were missed.

3.2 Multiple supply voltages

Chabini et al. [16] in their study proposed to minimize the dynamic power consumption under performance constraints by scaling down the supply voltage of computational elements off critical paths. In their work, the mixed-integer linear programming (MILP) method was first used to determine a schedule of the computational blocks that will lead to the maximum reduction of dynamic power consumption with designed performance constraints. Once the valid periodic schedule was computed, registers were inserted into the circuit to preserve the behavior of the original circuit. When compared to the design using the highest supply voltages, power reduction factors as high as 69.75% were obtained from their work.

Since multiple voltage domains are involved, multiple power grid structures are required. Power and floor plannings must, therefore, be implemented with care when employing this approach to optimize power consumption.

3.3 Clock gating for clock tree

According to Donno et al. [17], the power drawn by the clock tree in advanced microchips tends to dominate. Hence, they introduced a clock-gating approach that could be adopted at the register transfer level (RTL) to reduce clock power. In the algorithm that they developed, a clock tree topology, which balanced the reduction in clock switching against clock and activation function capacitive loading estimates, was first built. Clock-gating logic was then incorporated into the tree, achieving a balance between its power consumption and the power on the gated clock sub-tree. The physical and functional information was taken as inputs to generate a clock netlist at the output. The netlist was then used to update the structural description of the design. The results show that the capability of their approach in power saving is 75% better than conventional clock-gating methods.

The work proposed in ref. [17] focused on the calculation of the active and idle time frames of different registers and inserting the clock gating logics into the netlist. Since the algorithm implemented in the work did not account for an optimized location for the gating logic, the registers may end up being placed far apart after the netlist is updated. A larger clock network size and higher power consumed by the clock tree may, therefore, ensue.

3.4 Downsizing gates

Gate sizes are proportional to the parasitic capacitances. By this means, dynamic effective capacitance C_{dyn} could be reduced when gates are downsized. By selectively downsizing the gates of a circuit, Aizik and Kolodny [18] demonstrated that dynamic and leakage energy dissipations can be reduced. Doing so, however, may lead to an increase in speed delay. This is to say that the operating speed of a circuit can be traded off in exchange for power reduction. The findings reveal that 25% of dynamic power can be saved for circuits in 32-nm technology when the delay constraint is relaxed by 5%.

3.5 Interconnect-power reduction

A circuit consumes switching power P_{switch} when the interconnection capacitances are charged and discharged. The analysis in ref. [19] showed that the interconnect power occupied more than half of the total dynamic power consumption $P_{dynamic}$, with 90% of it contributed from 10% of the interconnections. To reduce C_{dyn}, larger wire spacing and minimal length routing were implemented for the high-power consuming interconnects. The researchers re-configured the interconnects without sacrificing the area and timing degradation and the results that they obtained showed that $P_{dynamic}$ was saved by 14%.

3.6 Clock network optimization

Lu et al. [20] demonstrated that power consumption is affected by the size of the clock network. They developed an algorithm that navigated the registers' locations during cell placement. When performing the navigation process, the Manhattan ring-based register guidance, the center of gravity constraints for registers, pseudo pin and net, and register cluster contraction were observed. The clock net wirelength was decreased by 16–33%, with no more than 0.5% increase in signal net wirelength.

However, precautions must be taken when adopting this approach, particularly for circuits with high densities. This is because, reducing the clock network size may increase the risk of routing congestion and this may lead to a poor signal net to wirelength.

3.7 Net ordering and wire space optimization

To optimize power consumption, Moiseev et al. [21] endeavored to reduce the capacitance for the most active nodes within parallel bundles. This can be achieved by finding the best arrangement of adjacent signals in the bundle and rearranging the positions of the wires so that the most active signal shares the smallest cross-capacitance. The approaches that they adopted are net ordering and wire space optimization. In their work, signals with high switching activity (*SA*) share a relatively larger space than those with lower *SA* (as seen in **Figure 7**). Further, the LSB is proposed to be

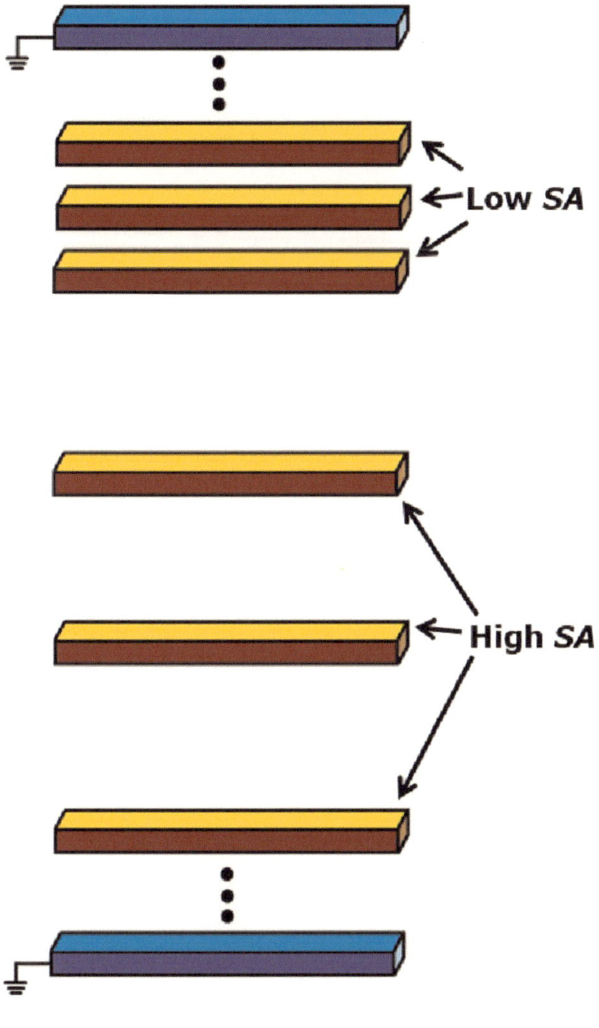

Figure 7.
Wires connected to signals with high SA share a relatively larger space than those with lower SA.

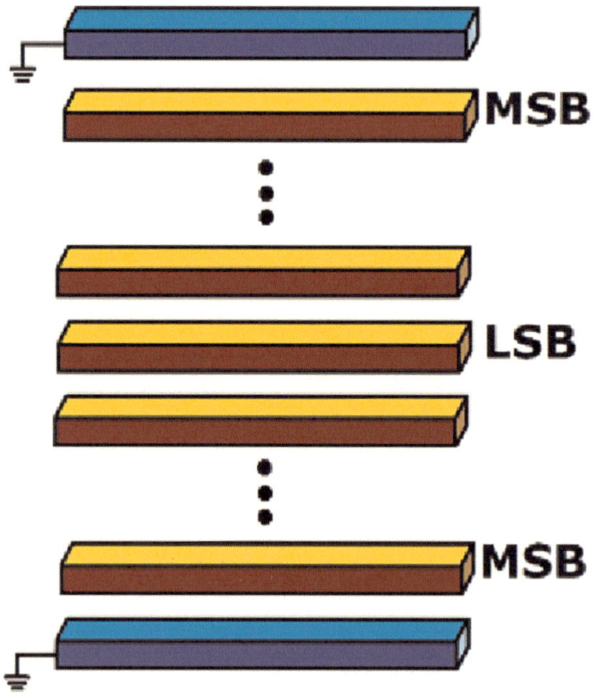

Figure 8.
Net ordering. The LSB (highest switching activity signal) is positioned at the center, while the MSB (lowest switching activity signal) is positioned near the sidewall.

placed at the center of the bundle and the MSBs at the ends, as depicted in **Figure 8**. This spacing and ordering optimization method was applied on industrial layouts of 65 nm process technology and the power saved ranged from 9 to 37%.

3.8 Leaving unused routing conductors floating

To reduce the effective coupling capacitance, Huda, Anderson, and Tamura [22] proposed to leave routing conductors adjacent to those used by timing critical or high activity nets floating. The purpose of doing so is to ensure that the original coupling capacitance among the conductors stays in series with other capacitances in the circuit. It is to be noted that, the equivalent capacitance of a chain of series capacitances is lower. As can be seen in **Figure 9**, tri-state buffers were used to disconnect the unused conductors. During high switching conditions, the tri-state buffer is allowed to be used as a normal buffer, without the loss or delay of data. The results show that the interconnect dynamic power was reduced up to approximately 15.5%, with a critical path degradation of about 1% and a total area overhead of about 2.1%.

4. Switching activity

The power optimization techniques discussed in the previous section involve redesigning the circuit topology. Most of these techniques are conducted based on the assessment of the switching activity. Switching activity *SA* comprises two basic

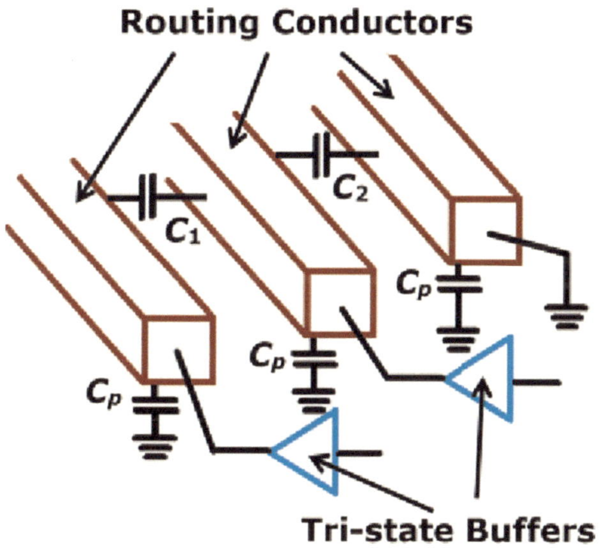

Figure 9.
The unused adjacent routing conductors are left floating, by connecting them to tri-state buffers. C_1 *and* C_2 *denote the coupling capacitance, while* C_p *denotes the plate capacitance (i.e., the parasitic capacitance formed between the substrate and the metal layers).*

elements, namely, (i) the toggle rate and (ii) static probability. The toggle rate gives an indication of the frequency the node toggles for a specified interval, and it is usually measured in millions of transistors per second. The static probability, on the other hand, predicts the logic state of the signal. An *SA* that shows 0.4 static probability, for example, suggests that the signal gives a logic 1 for 40% of the time and a logic 0 for the remaining 60%. **Figures 10** and **11** illustrate two sets of signals with opposite scenarios—the signals in **Figure 10** consist of identical toggle rates but different static probabilities, whereas those in **Figure 11** show different toggle rates, but the same static probability.

Switching activity is an important reference tool when performing power analysis. By reading the *SA*, critical design blocks that consume high power can be identified. The details of the circuit, such as where and when it has the lowest and highest toggle rate, can also be ascertained. This information is important in deciding the appropriate approach to save power.

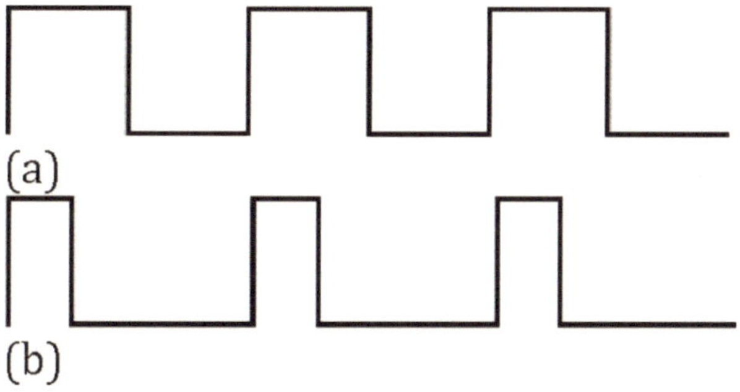

Figure 10.
The pulse trains in (a) and (b) consist of identical toggle rate but different static probabilities. For the pulse train in (a), toggle rate = 6 and static probability = 0.5; and that in (b), toggle rate = 6 and static probability = 0.25.

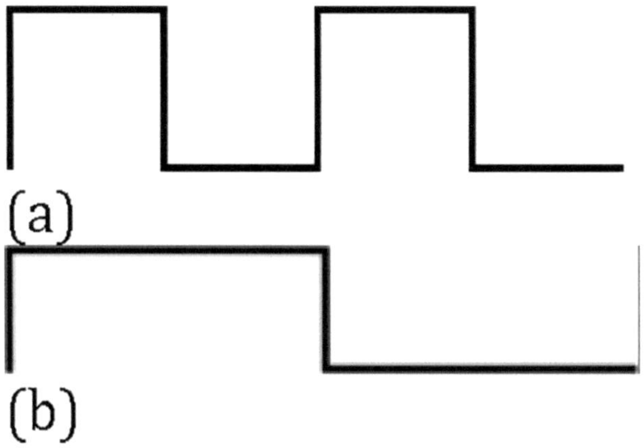

Figure 11.
The pulse trains in (a) and (b) consist of identical static probability but different toggle rates. For the pulse train in (a), toggle rate = 4 and static probability = 0.5; and that in (b), toggle rate = 2 and static probability = 0.5.

5. Conclusion

In this chapter, the power consumed by CMOS circuits is expounded. The total power consumption P_{total} in a CMOS circuit can be classified into two types—viz, the dynamic power $P_{dynamic}$ and static power P_{static}. Dynamic power refers to the power dissipated by the circuit when it is operating. It is induced by the switching activities, which take place at the nodes, and the short-circuit current formed at the transition state of the logic switch. Static power, on the other hand, occurs when the circuit is idle. It is caused mainly by the subthreshold and gate leakage currents. Since dynamic power takes up a significant fraction of the overall power consumption, different approaches have been developed to minimize it. The approaches focus on the reduction of the supply voltages, clock frequencies, or dynamic effective capacitance. By studying the activity factors of the design modules, the approaches can be applied to those with high power consumption.

Author details

Len Luet Ng[1], Kim Ho Yeap[2]*, Magdalene Wan Ching Goh[3] and Veerendra Dakulagi[4]

1 Intel Microelectronics, Bayan Lepas, Penang, Malaysia

2 Universiti Tunku Abdul Rahman, Kampar, Perak, Malaysia

3 Wawasan Open University, Penang, Malaysia

4 Guru Nanak Dev Engineering College, Bidar, Karnataka, India

*Address all correspondence to: yeapkh@utar.edu.my

References

[1] Yeap KH, Nisar H. Introductory chapter: VLSI. In: Yeap KH, Nisar H, editors. Very-Large-Scale-Integration. London, UK: InTechOpen; 2018. pp. 3-11

[2] Yeap KH, Nisar H. Introductory chapter: Complementary metal oxide semiconductor (CMOS). In: Yeap KH, Nisar H, editors. Complementary Metal Oxide Semiconductor. London, UK: InTechOpen; 2018. pp. 3-7

[3] Yeap KH, Isa MM, Loh SH. Introductory chapter: Integrated circuit Chip. In: Yeap KH, Hoyos JJS, editors. Integrated Circuits/Microchips. London, UK: InTechOpen; 2020. pp. 1-13

[4] Ho YK, Ahmad I, Sulong MS. Characterization of a 0.14 μm submicron NMOS with Silvaco TCAD simulator. Journal of Science and Technology. 2009;**1**:1-10

[5] Sulong MS, Ahmad I, Foo LT, Ho YK. Characterization of a 130 nm CMOS device using CVIV and focused ion beam. Jurnal Sains dan Teknologi. 2007; **4**:37-46

[6] Ho YK, Meng MK, Chun LK, Chiong TP, Nisar H, Rizman ZI. Design and analysis of 15 nm MOSFETs. Journal of Telecommunication, Electronic and Computer Engineering. 2016;**8**:1-4

[7] Yeap KH, Thee KW, Lai KC, Nisar H, Krishnan KC. VLSI circuit optimization for 8051 MCU. International Journal of Technology. 2018;**9**:142-149

[8] Rabaey JM, Chandrakasan A, Nikolic B. Digital Integrated Circuits: A Design Perspective. UK: Pearson Education; 2003

[9] Weste NH, Harris D. CMOS VLSI Design: A Circuits and Systems Perspective. India: Pearson Education; 2015

[10] Yeap KH. Fundamentals of Digital Integrated Circuit Design. 1st ed. UK: Author House; 2011

[11] Yeap KH, Lee JY, Yeo WL, Nisar H, Loh SH. Design and characterization of a 10 nm FinFET. Malaysian Journal of Fundamental and Applied Sciences. 2019;**15**:609-612

[12] Jacob B, Wang D, Ng S. Memory Systems: Cache, DRAM, Disk. US: Morgan Kaufmann; 2010

[13] Sharroush SM. An MTCMOS subthreshold-leakage reduction algorithm. In: 2nd Novel Intelligent and Leading Emerging Sciences Conference. Giza, Egypt: IEEE; 2020. pp. 7-14

[14] Chinta V. Subthreshold and Gate Leakage Current Analysis and Reduction in VLSI Circuits. Master of Science Thesis. Rochester, NY: Kate Gleason College of Engineering; 2007

[15] Weissel A, Bellosa F. Process cruise control: Event-driven clock scaling for dynamic power management. In: Proceedings of the International Conference on Compilers, Architecture, and Synthesis for Embedded Systems. Grenoble, France: Association for Computing Machinery (ACM); 2002. pp. 238-246

[16] Chabini N, Chabini I, Aboulhamid EM, Savaria Y. Methods for minimizing dynamic power consumption in synchronous designs with multiple supply voltages. IEEE Transactions on Computer-Aided Design of Integrated Circuits and Systems. 2003;**22**:346-351

[17] Donno M, Ivaldi A, Benini L, Macii E. Clock-tree power optimization based on RTL clock-gating. In: Proceedings of the 40th Annual Design Automation Conference. Anaheim CA USA: Association for Computing Machinery (ACM); 2003. pp. 622-627

[18] Aizik Y, Kolodny A. Finding the energy efficient curve: Gate sizing for minimum power under delay constraints. VLSI Design. 2011;**2011**: 845957

[19] Magen N, Kolodny A, Weiser U, Shamir N. Interconnect-power dissipation in a microprocessor. In: Proceedings of the International Workshop on System Level Interconnect Prediction. NY USA: Association for Computing Machinery (ACM); 2004. pp. 7-13

[20] Lu Y, Sze CN, Hong X, Zhou Q, Cai Y, Huang L, et al. Navigating registers in placement for clock network minimization. In: Proceedings of the 42nd Annual Design Automation Conference. NY USA: Association for Computing Machinery (ACM); 2005. pp. 176-181

[21] Moiseev K, Kolodny A, Wimer S. Timing-aware power-optimal ordering of signals. ACM Transactions on Design Automation of Electronic Systems. 2008; **13**:1-17

[22] Huda S, Anderson J, Tamura H. Optimizing effective interconnect capacitance for FPGA power reduction. In: Proceedings of the ACM/SIGDA International Symposium on Field-Programmable Gate Arrays; NY USA: Association for Computing Machinery (ACM); 2014. pp. 11-20

Chapter 6

Low-Power CMOS/FinFETs Circuit Using Adiabatic Switching Principle

Cancio Monteiro

Abstract

Power consumption has become a very serious concern with regard to the rapid technology of Internet of Things (IoT) devices. The IoT devices, such as sensor nodes, secure cryptographic devices, and medical implantable devices are general embedded systems that require low power and operate at low-frequency speed. Countless efforts have been done to reduce power consumption in complementary metal oxide semiconductors (CMOS) through supply voltage downscaling, reducing unnecessary clock activity, avoiding long path circuit topology, etc. Another circuit technique for low-power purpose is by employing adiabatic switching principle. The adiabatic switching is commonly used in minimizing energy loss during charging/discharging period at all nodes of the circuit. In this paper, a low-power adiabatic CMOS/FinFETs circuit for low-power secure logic application is presented. The circuit speed, power consumption, and other evaluation metrics indicating the circuit performances will be compared among the proposed circuits and other circuit topologies that are available in the literature.

Keywords: CMOS, adiabatic, low-power, FinFETs, dual-rail, PUF, secure logic, LSI multiplier

1. Introduction

In recent years, the emerging Internet of Things (IoT) technology has introduced challenges and opportunities for engineering-related fields. It is estimated that the number of active IoT devices will surpass 25.4 billion in 2030 [1], including wired and wireless sensor networks. Most researchers consider the security profile (authenticity, integrity, and confidentiality) [2–8] and power-saving crypto-devices [9, 10] as challenging efforts in IoT network design for resilient and sustainable infrastructure of Industry 4.0 [11]. With the rapid growth of portable and standalone IoT devices, the energy availability has to be well-managed to assure the sustainability of IoT connectivity. These IoT devices can be supplied either by utilizing abundant ambient energy sources [12] or by powering with rechargeable battery technology. In this context, the electronic circuit design technique that is able to consume low power has to be addressed. To contribute to the secure communication among IoT devices, the circuit designers are again demanded to produce secure cryptographic devices to withstand side-channel-analysis (SCA) attract techniques [13–16]. In tackling both the low-power and high-security demand, numerous efforts have been done at the circuit design level by employing the adiabatic switching principle [17]; such as secure adiabatic logic (SAL) [18], symmetric adiabatic logic (SyAL) [19], 2N-2N2P [20, 21], charge-sharing symmetric adiabatic logic (CSSAL) [22], 2-phase symmetric pass gate adiabatic logic (2-SPGAL) [23], and the secure quasi-adiabatic logic (SQAL) [24]. Moreover, to confirm the authenticity of any crypto-device, a physically unclonable function (PUF) circuit is utilized to verify the chip authenticity and for secure key generation [25, 26]. Definition of a PUF in [27] states that a PUF is a hardware security fundamental that translates an input challenge into an output response through a physical system in a manner that is specific to the exact hardware instance (unique) and cannot be replicated (unclonable). The PUF related SRAM-based circuit design in adiabatic operation was first reported in Quasi-Adiabatic Logic PUF (QUAL-PUF) [28]. Accordingly, the author of this paper then proposed the CMOS-based two-phase clocking adiabatic PUF (TPCA-PUF) [29], and the PUF circuit stability is further investigated under various temperature and process variations using FinFETS technology [30].

In this paper, the author further describes the adiabatic circuit design technique for low-power application, using single-rail and dual-rail circuit topologies. The proposed circuits' operation, the evaluation metrics utilized for secure logic verification, the frequency spectrum of the proposed circuits, and the LSI circuit design using proposed circuits in comparison with previous works to validate the effectiveness and the performances of the proposed works are presented.

The rest of this paper is structured as follows: Section 2 describes the fundamental low-power circuit design, which briefly describes the adiabatic switching principles in comparison with the conventional CMOS logic circuit. Section 3 presents the proposed CMOS logic circuit topologies in detail. Section 4 describes the proposed LSI circuits, their respective simulation conditions, and the security evaluation metrics. Simulation results and technical discussion of the proposed works in comparison with the convention-related circuits are discussed in section 5. Finally, Section 6 concludes the research findings of this work.

2. Low-power circuit design technique

To ensure the long battery life for battery-powered embedded cryptographic devices, the CMOS power consumption needs to be highly considered. There have been several circuit design techniques reported to reduce dynamic power consumption, such as reducing supply voltage to near and subthreshold regions, reducing circuit switching activities, and avoiding long critical paths to diminish unnecessary glitch current, etc. From the circuit supply voltage point of view, adiabatic logic principle is a promising technique that can guarantee the efficiency of power usage. Therefore, in the following subsections, the author describes power consumption comparison among conventional and adiabatic CMOS logic styles.

2.1 Power consumption of CMOS circuit

Total power in a CMOS circuit comes from dynamic power, short-circuit power, and static (or leakage) power, as indicated in **Figure 1**. The dynamic power consumption occurs when the output node's capacitor C_L is switched (charging period).

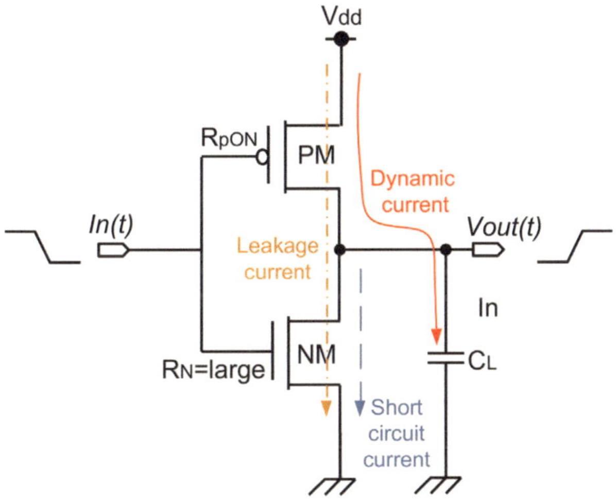

Figure 1.
Total power-on CMOS inverter: Dynamic Power, short-circuit power, and the leakage power.

The short-circuit power happens when both PMOS (PM) and NMOS (NM) transistors operate simultaneously during a short period of time of different input signal transitions (such as *In(t)* signal changes from 0 ➔ 1 and 1 ➔ 0). The other contributing power is the static power, which is consumed at either PM or NM transistor that operate in the cutoff region (or any electronic device is in standby mode).

$$P_{Total} = P_{Dynamic} + P_{SC} + P_{Static} \tag{1}$$

2.1.1 Dynamic power

Dynamic power consumption commonly depends on the switching frequency f, amplitude of power supply V_{dd}, and the load capacitance C_L of the output node. The operation of CMOS inverter logic in **Figure 1** is that when the state of input signal *In(t)* changes from 1 ➔ 0, the PM transistor is switched ON, and the current supply from V_{dd} is flowing down to charge the output node of C_L from initial condition of Vy(0_) = 0 ➔ Vy = V_{dd}. The internal equivalent RC model during this operation is called a pull-up network (PUN), as shown in **Figure 2a**. On the other hand, when the state of input signal *In(t)* charges from 0 ➔ 1, the NM transistor is switched ON and the output node of Vy is discharged from initial condition of Vy(0_) = V_{dd} ➔ Vy = 0 level (grounded). The internal equivalent RC model during this operation is called a pull-down network (PDN), as shown in **Figure 2b**.

From **Figure 2**, the total power dissipation can be calculated using each network system. By considering the MOS resistance value of $1/gm_n = 1/gm_p = R$, $C_L = C$, we can calculate the current source that flows into the circuit, as shown in Eq. (2):

$$ip(t) = i(t) = \frac{V_{dd}}{R} e^{\frac{-1}{RC}\tau} \tag{2}$$

The power consumption is calculated as:

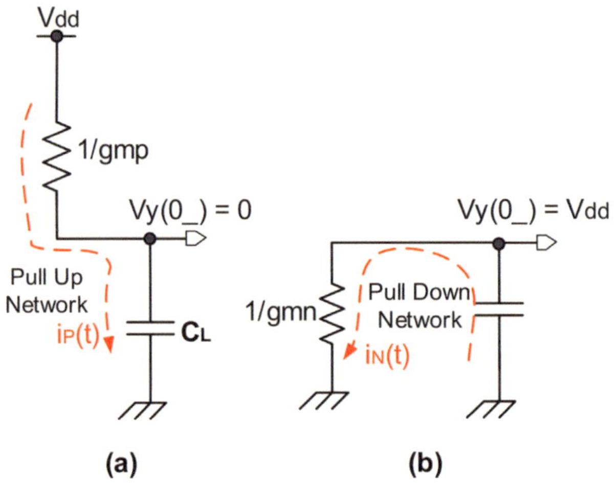

Figure 2.
(a) A CMOS pull-up network (PUN) RC equivalent model for charging phase, (b) A CMOS pull-down network (PDN) RC equivalent model for discharging phase.

$$p(t) = i(t)V_R(t) = i(t)^2R = \frac{{V_{dd}}^2}{R}e^{-2\frac{1}{RC}\tau} \tag{3}$$

Hence, the energy dissipated over the period of $t=0$ to $t=\tau$ is calculated as follows:

$$E_{charge} = \int_0^{\tau} p(t)dt = \int_0^{\tau} \frac{{V_{dd}}^2}{R}e^{-2\frac{1}{RC}\tau}dt = C\frac{{V_{dd}}^2}{2}\left(e^{-2\frac{1}{RC}\tau} + 1\right), \tag{4}$$

If, $\tau >> RC$, then the energy charged in output load capacitance is:

$$E_{charge} = \frac{1}{2}C{V_{dd}}^2 \tag{5}$$

From Eq. (5), half of the energy is dissipated as heat by the resistance $1/g_{mp}$ in **Figure 2a**; therefore, the total energy dissipated from power supply during PUN operation is $E_{total} = C{V_{dd}}^2$. Then, the average dynamic power $P_{dynamic}=E_{total}/T$, which is consumed during a certain period of time T can be formulated as

$$P_{dynamic} = \alpha f C{V_{dd}}^2, \tag{6}$$

where, the f denotes the clock frequency, and α is the switching activity factor, which corresponds to the average number of 0 ➔ 1 transitions that occur at the output cell in each clock cycle.

2.1.2 Short-circuit power

Short circuit power (P_{SC}) usually occurs because there is no zero second exist during different data transitions in CMOS logic circuit. The detailed discussion of short-circuit power was reported in [31], with an expression shown in Eq. (7);

$$P_{SC} = \frac{1}{12}\beta\tau f(V_{dd} - 2V_T)^3 \tag{7}$$

where β is a gain factor of a MOS transistor, τ represents the rise and fall time, f denotes a clock frequency, and the V_T is the MOS transistor threshold voltage.

2.1.3 Static power

Static power consumption is power loss when the transistor is not in the process of switching (cut-off state). It occurs when a small leakage current (I_{leak}) is flowing through the MOS transistor that is turned off. Static power is increasing significantly proportional to the shrinking of CMOS process technology.

There are several components that trigger the occurrence of leakage power [32, 33] as shown in **Figure 3**; such as (1) Reverse bias diode leakage current (I_{rbdl}), which occurs due to the reverse bias current of p-n junction between diffusion region of the transistor and substrate; (2) Gate oxide tunneling current (I_{ox}) is the leak current that flows from oxide insulation to substrate; (3) Gate induced drain leakage ($GIDL$) is another leakage current that increases exponentially due to the reduced gate oxide thickness; and (4) Subthreshold leakage current (I_{sub}). Thereby, the total summation of all leakage current I_{leak} components aforementioned can be formulated as:

$$P_{leak} = I_{leak} V_{dd} \tag{8}$$

2.2 Adiabatic switching principle

The adiabatic switching technique enables the logic circuit to reuse energy stored in output load capacitance during the recovery phase, known as energy recycling [17]. For better understanding of the adiabatic switching principle, the author uses the same RC model circuit with a different power supply as depicted in **Figure 4**. **Figure 4a** represents conventional logic with constant step V_{dd} voltage, whereas **Figure 4b** explains the concept of adiabatic switching with ramped step voltage, which is defined by the length of time.

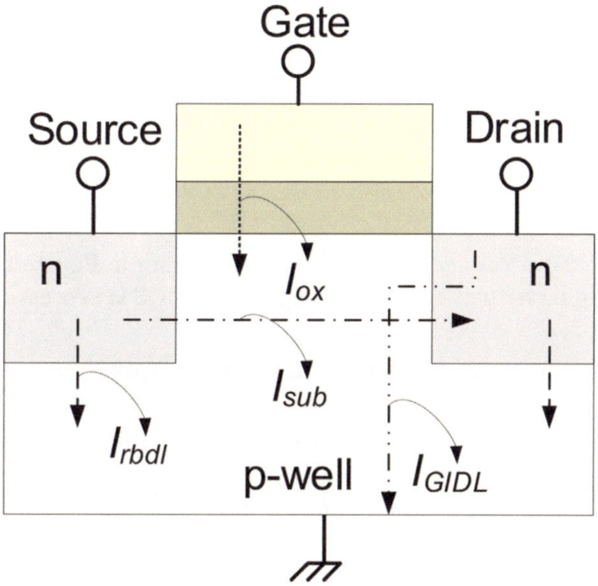

Figure 3.
Components of leakage power in CMOS [34].

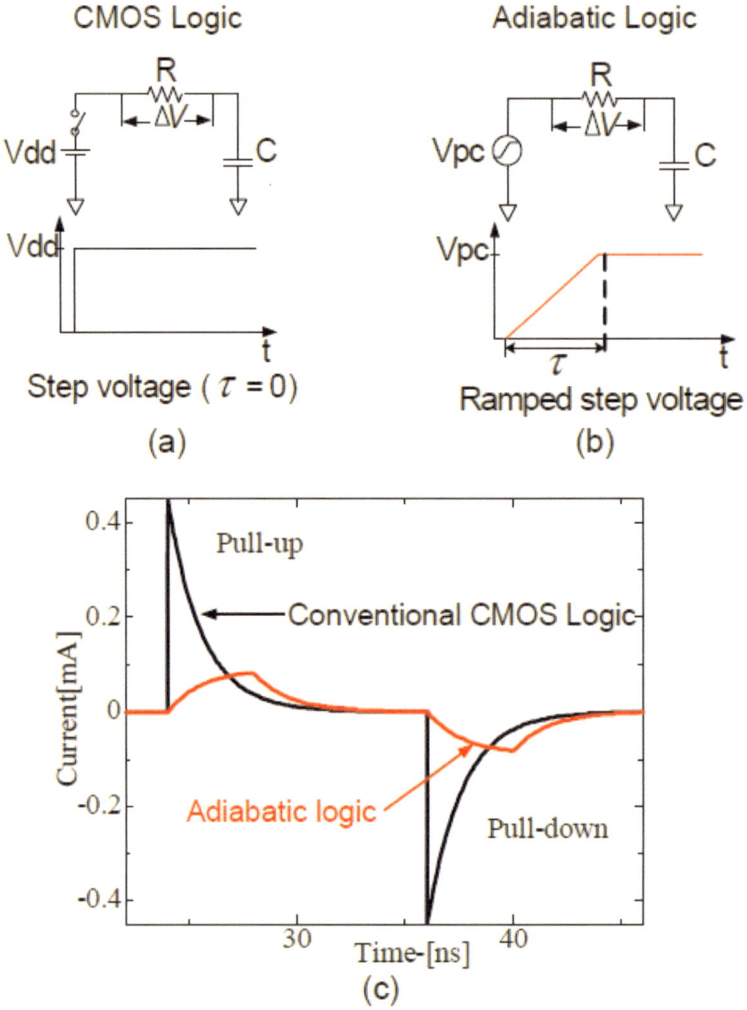

Figure 4.
Equivalent RC model of CMOS logic versus adiabatic logic; (a) CMOS logic with step voltage and (b) Adiabatic logic with ramped step voltage. (c) The peak supply current of the adiabatic logic is significantly lower than that of the conventional CMOS logic under the same simulation parameters and conditions.

Applying Kirchhoff Voltage Law (KVL) for the circuit in **Figure 1a** and **b**, the equation for charging network of the conventional CMOS is expressed in Eq. (9)

$$R_{i(t)} + \frac{1}{C}\int_0^T i(t)dt + v(0_-) = V_{dd} \tag{9}$$

and the charging network for adiabatic switching is similarly expressed in Eq. (10)

$$R_{i(t)} + \frac{1}{C}\int_0^T i(t)dt + v(0_-) = \frac{Vdd}{\tau}(t) \tag{10}$$

where τ is the rising time of ramp voltage *Vdd.* Applying the Laplace transform and inverse Laplace transform, we obtain the charging current as expressed in equations (11, 12) for CMOS logic and adiabatic logic, respectively:

$$i(t) = \frac{V_{dd}}{R} e^{-\frac{1}{RC}t} \tag{11}$$

$$i(t) = \frac{V_{dd}C}{\tau}\left(1 - e^{-\frac{1}{RC}t}\right) \tag{12}$$

The peak current difference of (Eqs. 11, 12) shows a large area and sudden flow of the current of the conventional CMOS, and gradual increase of supply current peak of the adiabatic switching in accordance with slow rising τ, which can be observed in **Figure 1c**.

Further analysis from an energy consumption perspective, the dissipated energy over the period $t = 0$ to $t = \tau$ is expressed as in Eq. (13)

$$E_{diss} = \int_0^{\tau} Ri^2(t)dt + E(0_-) \tag{13}$$

Substituting current $i(t)$ in Eqs. (11, 12) into Eq. (13), we have energy stored in capacitance for each conventional CMOS and adiabatic switching as expressed in Eqs. (14, 15), respectively.

$$E_{CMOS} = \frac{1}{2} CV_{dd}^2 \tag{14}$$

$$E_{Adiabatic} = \frac{RC}{\tau} CV_{dd}^2 \tag{15}$$

Eq. (15) obviously shows that by increasing the time of τ, the energy dissipation of adiabatic logic is significantly lower compared to the one of the conventional CMOS logics in Eq. (14).

3. CMOS logic circuit topology

The logic circuit available in the literature has two kinds of circuit topologies; the single-rail (SR) logic circuit composes of static CMOS (scCMOS: see **Figure 1**) and dynamic CMOS logics [35], and the dual-rail CMOS logic (DR-CMOS or differential logic) [20, 36], as depicted in **Figure 5a** and **b**. Regarding these circuits, uncountable research have been done from the viewpoint of low-power dissipation [20, 36–41], high speed, and further application into the secure cryptographic hardware design [18–24, 42, 43]. From the logic's security perspective, balancing supply current flows into the circuit is the main constraint, since the side-channel cryptanalysis targeting for the different peak current/power traces when crypto devices execute encryption and decryption processes [14]. Hence, **Figure 5** describes the supply current traces at different input data transitions for conventional static CMOS, dual-rail CMOS circuits (refers to **Figure 5a** and **b**, respectively), and our previously proposed charge-sharing symmetric adiabatic logic (CSSAL [22]). Effective side-channel analysis countermeasure is how the circuit is able to mask different input transitions with the same supply peak current despite any input–output data flipping. This can be solved by the charge-sharing technique of the proposed CSSAL circuit. In addition, the CSSAL adopted the adiabatic switching principle, which lower peak current compared with the conventional CMOS logic technique in scCMOS and DR-CMOS in **Figure 5**.

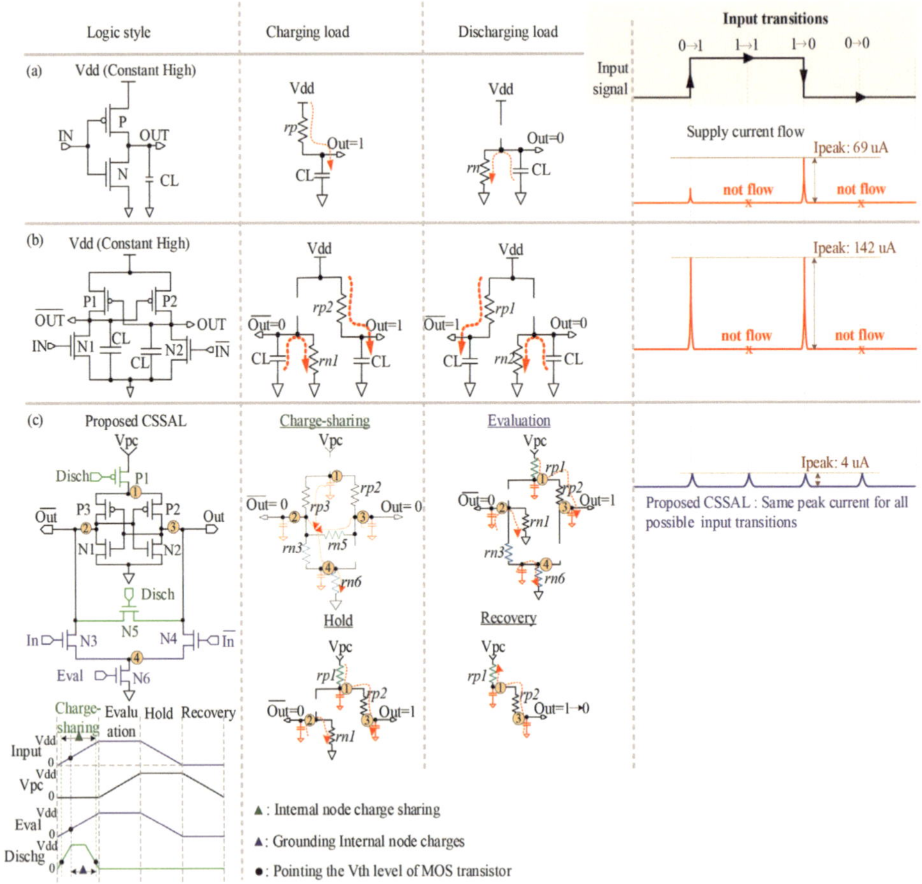

Figure 5.
Logic circuit topology and each of its supply current traces.

4. Proposed LSI circuit

The author of this paper has proposed three different circuit applications based on SR and DR CMOS circuit topologies for low-power and high-security profile, such as CSSAL [22], the source biased semi-adiabatic logic (SBSAL) [41], and the two-phase clocking adiabatic physical unclonable function (TPCA-PUF) [29]. In these following sub-sections, the author will present the fundamental circuit topology of each and their respective LSI block diagram.

4.1 The CSSAL circuit

The fundamental inverter logic style of the CSSAL circuit is shown in **Figure 5c**. The CSSAL is designed using DR circuit topology with four phases of adiabatic switching operations (charge-sharing, evaluation, hold, and recovery phases) [22], in which, the same internal equivalent RC model of each phase occurs for all possible different input data transitions, which yielding the same peak current as depicted in the right side of **Figure 5c**. It is obviously shown in **Figure 5** that the CSSAL performs balanced low peak current in comparison with the other logic circuits along the four different input transitions. This type of supply current trace is difficult to predict the position of its true input data, hence it is secure and applicable for cryptographic LSI design (**Figure 6**). To validate the security merit of the proposed CSSAL, the bit parallel cellular multiplier over finite field $GF(2^4)$ has been designed and implemented using the 0.18 um CMOS process technology. Input-output signals of the bit parallel cellular multiplier over $GF\ (2^4)$ are depicted in **Figure 7**.

4.2 The SBSAL circuit

The proposed SBASL circuit is a type of SR static CMOS logic family in adiabatic switching operation with sinusoidal power clock supply, as depicted in **Figure 8d**. The SBSAL circuit is basically operated in charging and discharging periods, in which the equivalent RC model of PUN and PDN are depicted in **Figure 9a** and **b**. This figure illustrates the output voltages, the instantaneous power, and the energy dissipated during charging and discharging phases. The total energy loss in SBSAL logic circuit is formulated in Eq. (16) as follows:

$$E_{SBSAL} = \frac{RC}{\tau} CV_{PC}{}^2 + \frac{1}{2} CV_{bias}{}^2 + \frac{RC}{\tau} C(V_{out} - V_{bias})^2 \tag{16}$$

This Eq. (16) means the energy stored in the load capacitance C_L is recycled to V_{bias} power supply. Although there is nonadiabatic energy loss of $\frac{1}{2}CV_{bias}{}^2$ in Eq. (8), the V_{bias} is set to 0.23 V, which has very low contribution to the total energy loss in the circuit. The SBSAL circuit PDN network is connected to 0.23 Volt bias voltage instead of connecting to ground or another sinusoidal supply voltage. This connection technique will only require one circuit to produce V_{pc} power supply. This means that the proposed SBSAL has low complexity if compared to the other adiabatic logic family shown in **Figure 8b** and **c**.

To validate the effectiveness of the proposed logic as a low-power SBSAL circuit, we implemented a 4x4-bit array SBSAL LSI multiplier as depicted in **Figure 10**. It is verified that the SBSAL multiplier logic function is well operated as shown in **Figure 11**.

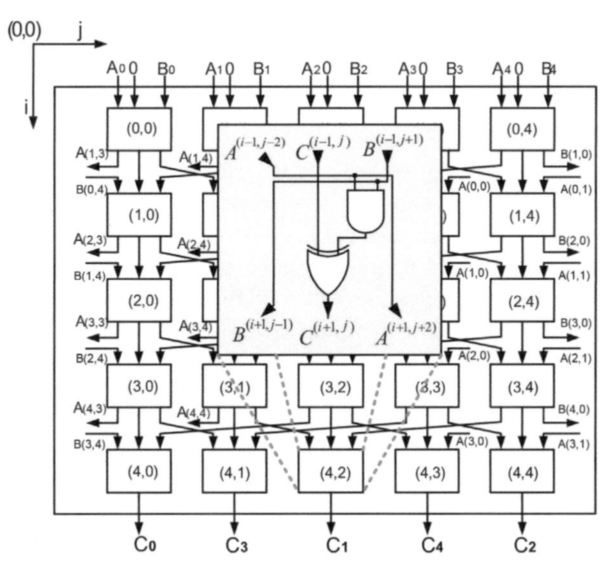

Figure 6.
The circuit block diagram of the bit parallel cellular multiplier over GF (2^4).

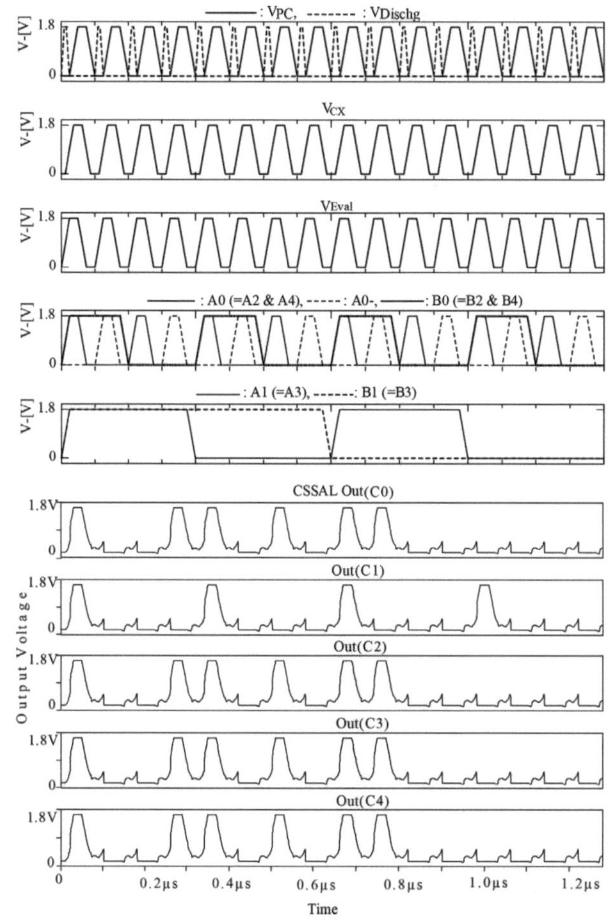

Figure 7.
Input-Output signals of the CSSAL bit parallel cellular multiplier over GF (2^4).

4.3 The TCPA-PUF circuit

The proposed adiabatic FinFETs based PUF circuit topology is depicted in **Figures 12** and **13a**. It was designed with cross-coupled latch circuit based on SRAM circuit topology. The challenge signal of the proposed PUF circuit is controlled by the static CMOS inverter aimed to conduct charging and discharging of the PUF cell semi-adiabatically using a trapezoidal power clock signal of Vpc. Notable improvement from QUAL-PUF circuit topology, the TCPA-PUF controls the current flow from output nodes to slowly flow to the ground through N4 transistor by controlling its operation speed with a ramped Vpc- signal. Notably, in the proposed adiabatic PUF circuit, the author applies two phases of power clock signals Vpc and Vpc-, as depicted in **Figure 14b**. The circuit operation of the TCPA-PUF cell is shown in **Figure 15**. Detailed TCPA-PUF circuit operation in an adiabatic mode for CMOS-based design has been clearly explained in [29], and the FinFETs-based TCPA-PUF design can be accessed in [30].

To verify the effectiveness and the stability of the proposed SRAM based TCPA-PUF, the author designs a 4-bit cascaded adiabatic PUF as depicted in **Figure 14**. Each local PUF is supplied with four different power clocks with a phase difference of 90.

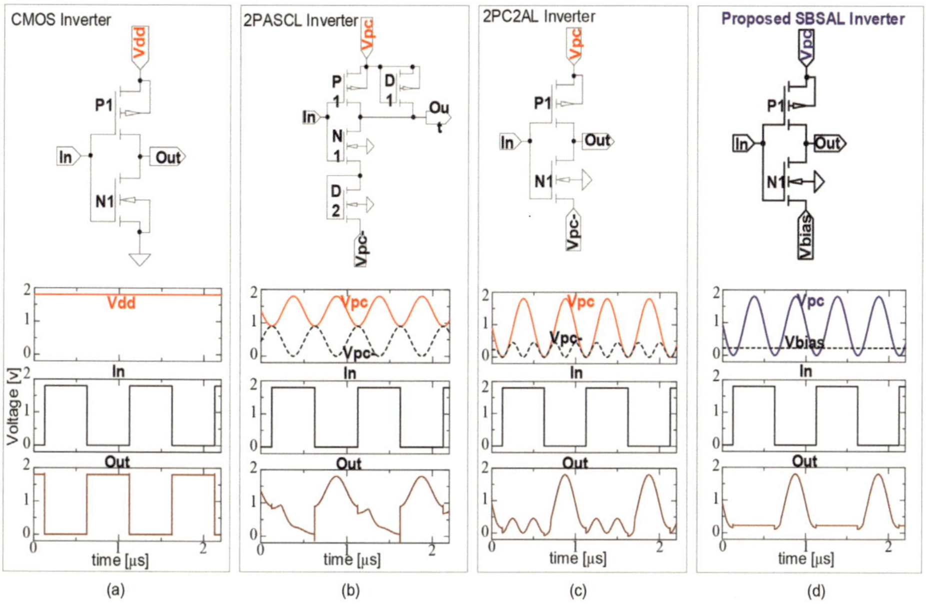

Figure 8.
Inverter logic investigated; a) Conventional CMOS logic, b) Adiabatic 2PASCL logic [39], c) Adiabatic 2PC2AL logic [40], and d) Proposed SBSAL [41].

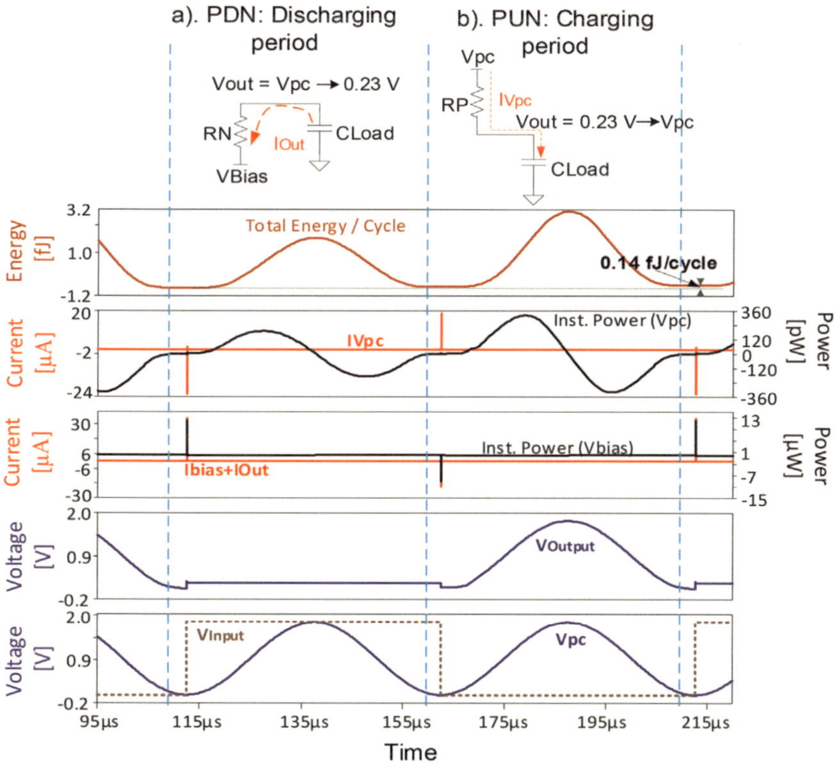

Figure 9.
Proposed SBSAL logic operation; a) Discharging period and b) Charging Period.

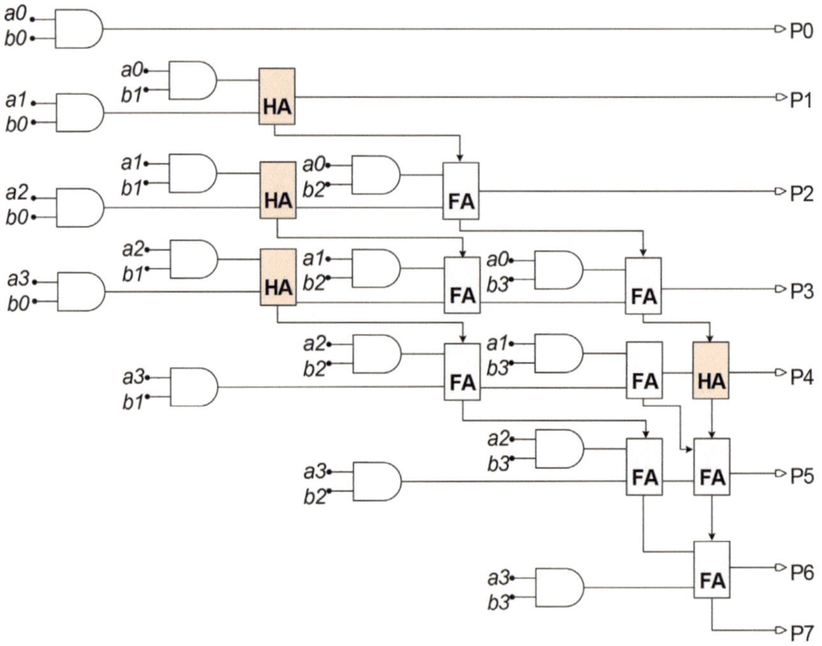

Figure 10.
Circuit diagram of a 4x4-bit array LSI multiplier.

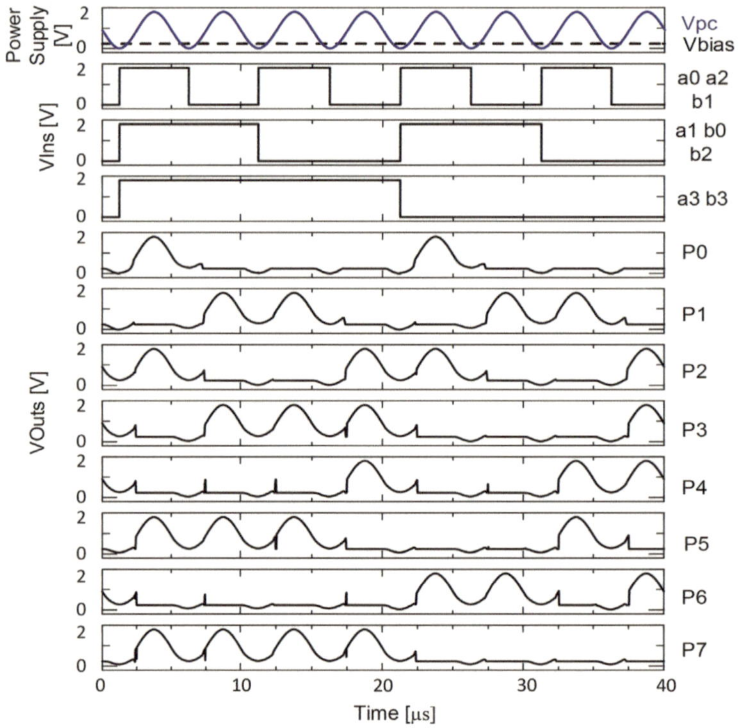

Figure 11.
Input-output signals of SBSAL multiplier at 25 MHz.

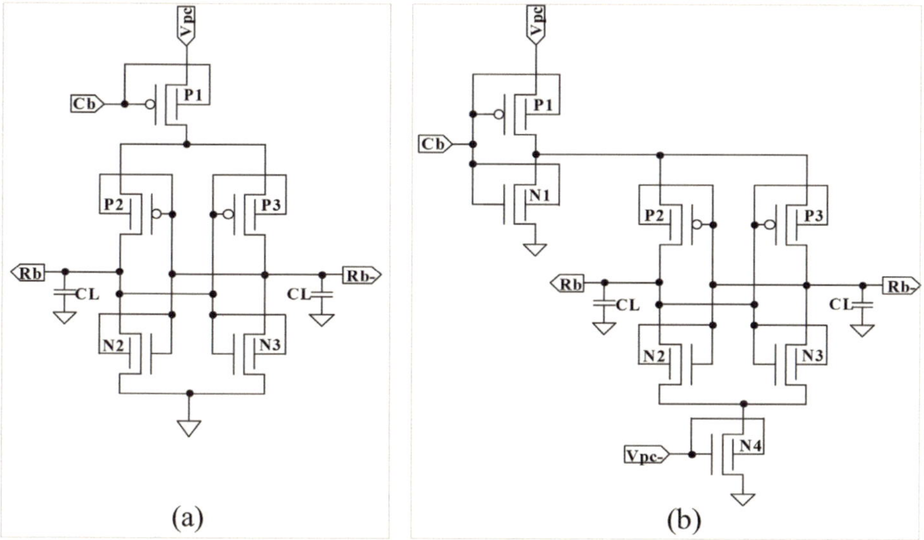

Figure 12.
SRAM-based FinFET PUF circuit; (a) QUAL-PUF circuit, (b) TCPA-PUF circuit.

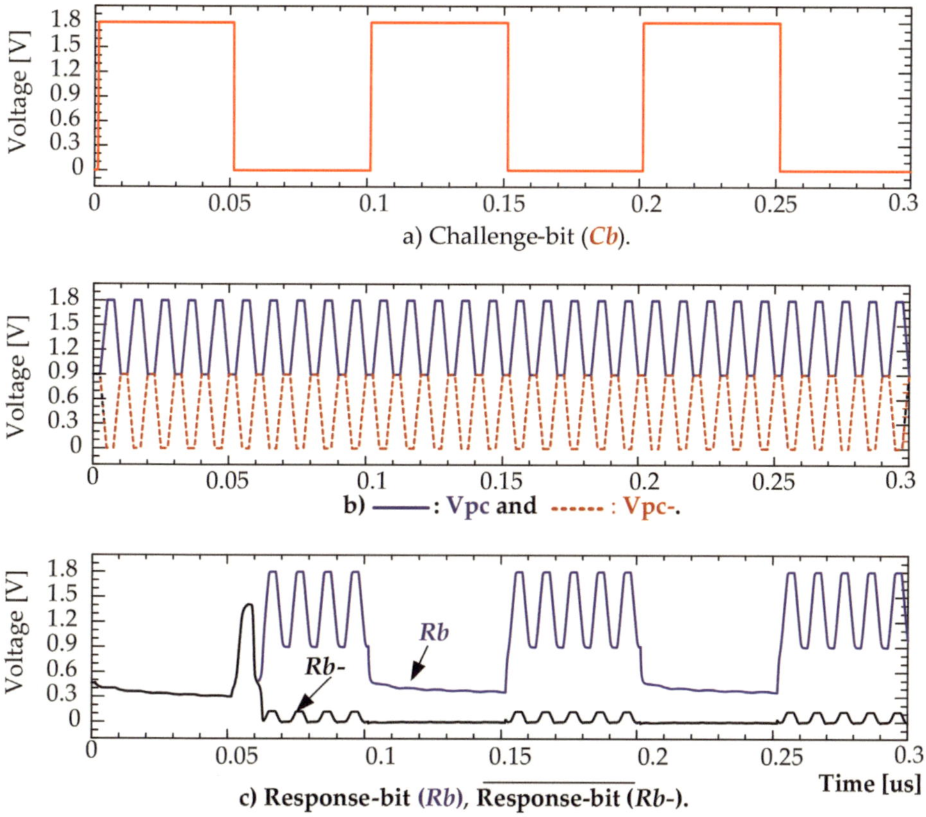

Figure 13.
Input and output signals of the proposed CMOS TCPA-PUF cell with nominal 1.8 V of Vdd voltage.

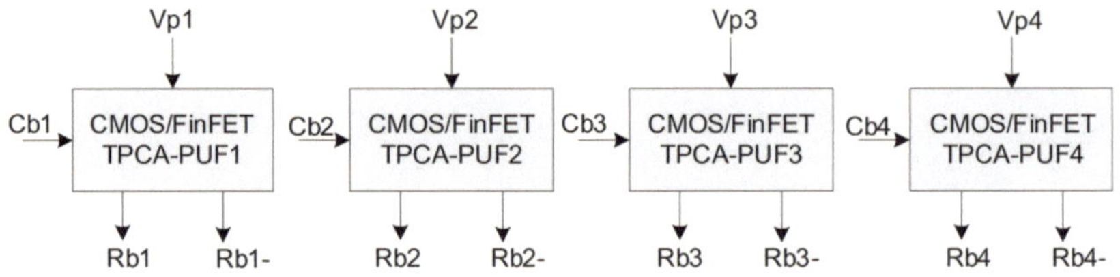

Figure 14.
Proposed 4-bit CMOS/FinFET TPCA-PUF architecture.

Moreover, each challenge bit signal has ¼ delay time of one power clock cycle. This delay time allows the challenge bits to flip the response signals right at the middle point of the idle/wait phase of the *Vpc* signals, and the challenge bits are perfectly flipped adiabatically.

Monte-Carlo simulation results of the 4-bit TPCA-PUF and QUAL-PUF challenge-response signals are depicted in **Figure 16**, where 100 times repetitions of the same 4-bit LSI PUF circuit are simulated. This result is performed with reference temperature of T = 27°C and C_L = 10 fF, f_{Cb} = 10 MHz, and f_{Vpc} = 100 MHz with ±10% of Vth variation. Simulation results of response signals (Rb1–Rb4) with a given challenge bit (Rb) performed correct and stable operations for both PUF circuit topologies.

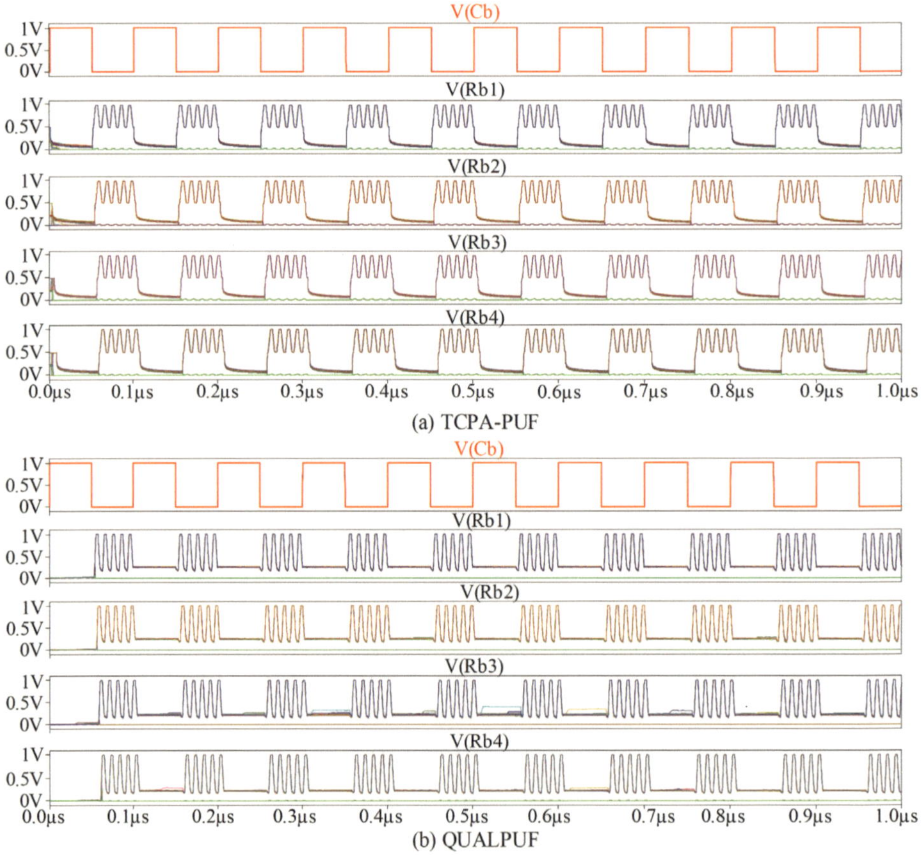

Figure 15.
Monte-Carlo simulation result of proposed 4-bit FinFET based TPCA-PUF LSI circuit with nominal 1 V of Vdd voltage.

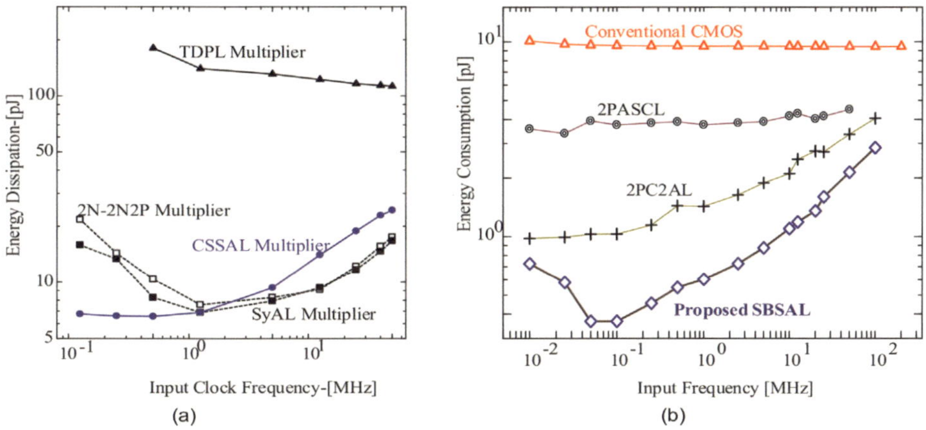

Figure 16.
Energy dissipation of the LSI multiplier circuits; (a) bit parallel cellular multiplier over GF (2^4) *with secure CSSAL circuit (refer to* **Figure 6***), (b) 4x4-bit array multiplier (refer to* **Figure 10** *with SBSAL circuit).*

5. Simulation results and discussion

The simulation results presented in this section are all obtained from LTSpice simulation of both CMOS and FinFETs technologies, while the simulation conditions are described in **Table 1**.

The technical discussion in this paper will only focus on energy dissipation, which reflects the title of low-power of this paper. Energy dissipated by each LSI circuits is obtained from the following Eq. (17) formula:

$$E_{SBSAL} = \int_0^T \left(V_{pc(t)} I_{pc(t)} + V_{bias} I_{bias(t)}\right) dt, \tag{17}$$

and energy dissipation for both CSSAL and TCPA-PUF are formulated in Eq. (18):

$$E_{diss.} = \int_0^T \Sigma(Vpcs.Ipcs) dt. \tag{18}$$

CSSAL and SBASL TCPA-PUF circuits	• Vpc Max.: 1.8 V with f_{Vpc}: 125 KHz–50 MHz (CSSAL Multiplier *GF* (2^4) LSI) • Vpc Max.: 1.8 V with f_{Vpc}: 10 KHz–100 MHz (SBSAL 4x4-bit Multiplier LSI) • *Vpc*: swing from 0.9–1.8 V for CMOS trapezoidal clock (FinFETs: 0.5–1 V), f_{Vpc} = 100 MHz • *Vpc-*: swing from 0–0.9 V for CMOS trapezoidal power clock (FinFETs: 0–0.5 V), f_{Vpc-} = 100 MHz • *Cb* voltage: 1.8 V CMOS pulse signal (FinFETs: 1 V), f_{Cb} = 10 MHz
Transistor parameter and ratio	• CMOS Parameter: 0.18 µm ROHM standard CMOS process with ratio W/L= 0.6 µm/ 0.18 µm for all NMOS and PMOS Transistors • FinFET Parameter: 45 nm with bulk, the ratio W/L = 60 nm/45 nm for all PMOS and NMOS Transistor

Table 1.
Simulation conditions.

1. In CSSAL design process, the author has employed several techniques, such as (1) adopting the adiabatic switching principle for energy recycling to achieve low power consumption and low peak current, (2) dual-rail logic circuit topology is utilized to establish uniform transitional supply peak current, and (3) symmetric pull-down network transistors with internal node charges are shared and discharged to ground simultaneously, which construct a constant internal equivalent RC model for all input condition to reduce current-to-data dependency. The evaluation metric in our proposed CSSAL circuit has two targets: the secure logic and low power. For secure logic verification, we evaluate the logic ability to balance current traces by calculating the normalized standard deviation as in following Eq. (19):

$$\mathrm{NSD} = \boldsymbol{\sigma}_{\mathrm{E}}/\bar{\mathrm{E}}, \tag{19}$$

where the $\bar{\mathrm{E}}$ is the average of energy dissipation of every respective input transition, and the standard deviation of $\boldsymbol{\sigma}_{\mathrm{E}} = \sqrt{\sum_{i=E1}^{En} (Ei - \bar{\mathrm{E}})^2/n}$. The ideal value of NSD has to be 0%. The post-layout comparison of secure logic circuits in this paper is as labeled in **Figure 16a**, such as TDPL, SyAL, CSSAL, and the 2N-2N2P. The NSD result calculated at 1.25 MHz has shown that the CSSAL has 3.49%, SyAL: 4.69%, 2N-2N2P: 49.08%, and TDPL has 58.71%. Moreover, the energy dissipation per cycle of post-layout simulation is shown in **Figure 16a**, in which the proposed CSSAL consumes low energy at lower frequencies (1.25 MHz and below). Therefore, the proposed CSSAL cellular multiplier is suitable for low-power and high-security devices at 1.25 MHz and/or below this speed.

2. The energy of SBSAL is checked and compared with other adiabatic static logic families including conventional static CMOS logic as depicted in **Figure 16b**. It is obviously shown in this figure, the proposed SBSAL multiplier has reduced energy about 94% from conventional CMOS circuit, 84% from the 2PASCL circuit, and 58% from the 2PCAL circuit at 1~MHz operating frequency, and always consumes lower energy along the frequency band investigated in this work.

3. The proposed TCPA-PUF circuit stability has been verified in the 180 nm CMOS process and in 45 nm bulked FinFETS technology, where the proposed circuit has performed its superiority in terms of evaluation metrics (*Uniqueness* and *Reliability)* and low-power consumption than that of the QUAL-PUF one. The ***Uniqueness*** is used to determine the ability of a PUF to uniquely distinguish a chip among the other chips [28], as formulated in the following Eq. (20):

$$Uniquness\ (U(\%)) = \frac{2}{k(k-1)} \sum_{i=1}^{k-1} \sum_{j=i+1}^{k} \frac{HD(R_i, R_j)}{n} \times 100. \tag{20}$$

The ***Reliability*** measures how reproducibly the challenge-response pairs of a PUF instance with the varying environmental conditions such as temperature and CMOS process variations as shown in Eq. (21):

$$Reliability\ (R(\%)) = 100 - \frac{1}{k} \sum_{i=1}^{k} \frac{HD(R_i, R_{i,j})}{n} \tag{21}$$

PUF	QUAL-PUF [28]			Proposed TCPA-PUF	
	CMOS		FinFET	CMOS	FinFET
Year	2020		2021	2021 [29]	2021 [30]
Tech.	180 nm	45 nm	45 nm	180 nm	45 nm
Topology	Adiabatic SRAM			Adiabatic SRAM	
Transistor-number/bit	5	5	5	7	7
Process (nm)	180	45	45	180	45
Start-Up power	3.08 μW	NA	65.69 nW	0.47 μW	18.32 nW
Energy (fJ/bit/cycle)	39.18	0.08	7.36	15.98	2.30
Uniqueness (%)	40.50	49.41	49.46	49.82	50.13
Reliability (%)	96.20	99.60	99.47	99.47	99.57
BER (%)	3.8	0.4	0.53	0.53	0.43

Table 2.
Comparison of conventional and proposed adiabatic PUFs (with T = 27°C and C_L = 10 fF, f_{Cb} = 10 MHz, and f_{Vpc} = 100 MHz).

The ideal values of *uniqueness* and *reliability* are 50% and 100%, respectively. The TCPA-PUF evaluation results have always been close to the ideal values.

It has been revealed that the FinFET device has several advantages, such as higher on-state current, lower off-state current (lower leakage current), faster-switching speed [44], and its double gates enabling three possible connection modes (shorted gate-SG, independent gate-IG, and low-power-LP) for low power and high-speed applications. In this work, the author has thoroughly investigated the proposed TPCA-PUF cell using bulked FinFET with a 45 nm process for all SG, IG, and LP modes. As a result, the author has revealed that SG mode is suitable for the proposed TPCA-PUF circuit topology. The gate connection type of LP and IG modes leads to higher energy and produces wrong response bits for larger cascaded bit-length (4-bit in this work). Therefore, the whole works of 4-bit LSI design and simulation, including the data presented in this paper are performed by utilizing the SG mode connection type. The TPCA-PUF cell was implemented using SRAM-based circuit topology, hence this study is claimed to be the first work in the literature that employs FinFETs-based SRAM type PUF. Numerical data in **Table 2** compare the QUAL-PUF and proposed TPCA-PUF for both CMOS and FinFETs process technologies. Overall data in **Table 2** have shown that the proposed TPCA-PUF consumes lower energy/bit/cycle and start-up power, which is suitable for low-power IoT application.

6. Conclusion

This paper has presented a comparative study on energy dissipation and secure evaluation metrics of the proposed CSSAL, SBSAL, and QUAL-PUF with other conventional related circuit topologies.

1. Secure CSSAL: the NSD result calculated at 1.25 MHz has shown that the CSSAL has 3.49%, SyAL: 4.69%, 2N-2N2P: 49.08%, and TDPL has 58.71%. Moreover, the energy dissipation per cycle of post-layout simulation has shown that the

CSSAL consumes low energy at lower frequencies (1.25 MHz and below). Therefore, the proposed CSSAL cellular multiplier is suitable for low-power and high-security devices at 1.25 MHz and/or below this speed.

2. Low-Power SBSAL: Simulation results have shown that the proposed SBSAL multiplier has reduced energy about 94% from conventional CMOS circuit, 84% from the 2PASCL circuit, and 58% from the 2PCAL circuit at 1~MHz operating frequency.

3. TCPA-PUF: the SRAM-based CMOS and FinFETs PUF using 180nm and 45 nm technology process, respectively, has been further investigated into 4-bit cascaded bitlength, where the proposed TPCA-PUF has reduced energy/bit/cycle and start-up power, both about 70% from the QUAL-PUF cell at the same reference temperature of 27°C.

The uniqueness, reliability, and the BER of the proposed FinFETs-based TPCA-PUF are 50.13%, 99.57%, and 0.54%, which exhibits a superior security performance if compared with the FinFETs-based QUAL-PUF cell. The remarkable performances (ultra-low power and security profile) of the proposed FinFETs-based TPCA-PUF makes it an appropriate candidate for low-power and secure IoT device applications.

Author details

Cancio Monteiro
Faculty of Engineering Science and Technology, Department of Electronics and Electrical Engineering, National University of Timor Lorosa'e (UNTL), Dili, Timor-Leste

*Address all correspondence to: cancio.monteiro@untl.edu.tl

References

[1] Bojan J. Internet of Things statistics for 2022 – Taking Things Apart. DataProt. 2022. Available online: https://dataprot.net/statistics/iot-statistics/ (update: May 13, 2022; Accessed: July 2, 2022)

[2] Atzori L, Iera A, Morabito G. The internet of things: A survey. Computer Networks. 2010;**54**:2787-2805

[3] Bandyopadhyay D, Sen J. Internet of things: Applications and challenges in technology and standardization. Wireless Personal Communications. 2011;**58**:49-69

[4] Keoh SL, Kumar SS, Tschofenig H. Securing the internet of things: A standardization perspective. IEEE Internet of Things Journal. 2014;**1**: 265-275

[5] Sicari S, Rizzardi A, Grieco L, Coen-Porisini A. Security, privacy and trust in internet of things: The road ahead. Computer Networks. 2015;**76**:146-164

[6] Radomirovic S. Towards a model for security and privacy in the internet of things. In: Proceedings of the First International Workshop on Security of the Internet of Things. Tokyo, Japan; 2010

[7] Wurm J, Hoang K, Arias O, Sadeghi AH, Jin Y. Security analysis on consumer and industrial iot devices. In: Proceedings of the 2016 21st Asia and South Pacific Design Automation Conference (ASP-DAC). Macao, China; 2016. pp. 519-524

[8] Babaei A, Schiele G. Spatial reconfigurable physical unclonable functions for the internet of things. In: Proceedings of the International Conference on Security, Privacy and Anonymity in Computation, Communication and Storage. Guangzhou, China; 2017. pp. 312-321

[9] Martinez B, Montón M, Vilajosana I, Prades JD. The power of models: Modeling power consumption for iot devices. IEEE Sensors Journal. 2015;**15**: 5777-5789

[10] Mukhopadhyay SC, Suryadevara NK. Internet of things: Challenges and opportunities. In: Internet of Things. Berlin/Heidelberg, Germany: Springer; 2014. pp. 1-17

[11] Shrouf F, Ordieres J, Miragliotta G. Smart factories in industry 4.0: A review of the concept and of energy management approached in production based on the internet of things paradigm. In: Proceedings of the 2014 IEEE International Conference on Industrial Engineering and Engineering Management (IEEM). Selangor, Malaysia; 2014. pp. 697-701

[12] Paul K. Timing attacks on implementation of Diffie-Hellman, RSA, DSS and other system. In: Proceedings of the 16th Annual International Cryptology Conference, Santa Barbara, California, USA. 1996. pp. 104-113

[13] Erick O, Rincón-Mora GA. Electrostatic energy-harvesting and battery-charging CMOS system prototype. IEEE Transactions on Circuits and Systems. 2009;**56**:1938-1948

[14] Kocher P, Jaffe J, Jun B. Differential power analysis. Proceedings of the International Advances in Cryptology Conference (CRYPTO). 1999:388-397

[15] Elke DM, Siddika BO, Bart P, Ingrid V. Differential electromagnetic attack on an FPGA implementation of elliptic

curve cryptosystems. In: Proceedings of World Automation Congress (WAC'06), Budapest, Hungary. 2006. pp. 1-6

[16] Eric B, Christophe C, Francis O. Correlation power analysis with a leakage model. In: Proc. Sixth Int. Workshop on CHES 2004, Cambridge, MA, USA: LNCS; 2004. pp. 16-29. DOI: 10.1007/978-3-540-28632-5-2

[17] Athas WC, Svesson LJ, Koller JG, Traztzanis N, Chuo EY-C. Low power digital system based on adiabatic-switching principles. IEEE Transactions on Very Large Scale Integration System. 1994;**2**:398-406

[18] Khatir M, Moradi A. Secure adiabatic logic: A low energy DPA resistant logic style. Cryptology ePrint Archive, 2008. https://ia.cr/2008/123

[19] Choi BD, Kim KE, Chung K-S, Kim D. Symmetric adiabatic logic circuits against differential power analysis. ETRI Journal. 2010;**32**(1):166-168

[20] Kramer A, Denker JS, Flower B, Moroney J. 2nd Order Adiabatic Computation 2N-2P and 2N-2N2P Logic Circuits. In: Proceedings of IEEE International Symposium on Low Power Design, Dana Point California, USA. 1995. pp. 191-196

[21] Moradi A, Khatir M, Salmasizadeh V, Shalmani MTM. Investigating the DPA-resistance property of charge recovery logics. IACR ePrint Archive. 2008, pp. 191—192

[22] Cancio M, Yasuhiro T, Toshikazu S. Charge–sharing symmetric adiabatic logic in countermeasure against power analysis attack at cell level. Microelectronics Journal. 2013;**44**:496-503

[23] Amit D, Himanshu T. 2-Phase adiabatic logic for low-energy and CPA-resistant implantable medical devices. IEEE Transactions on Consumer Electronics. 2022;**68**(1):47-56

[24] Moshe A, Hadar D, Itamar L, Osnat K, Alexander F. DPA-secured quasi-adiabatic logic (SQAL) for low-power passive RFID tags employing s-boxes. IEEE Transactions on Circuits and Systems. 2015;**62**(1): 149-156

[25] Charles H, Meng-day Y, Farinaz K, Srinivas D. Physical unclonable functions and applications: A tutorial. Proceedings of the IEEE. 2014;**102**(8): 1126-1141

[26] Debdeep M. PUFs as promising tools for security in Internet of Things. IEEE Design & Test. 2016;**33**(3):103-115

[27] Thomas MG, Ibrahim EB, Zhiming MW, Utz R, Robert JY. A PUF taxonomy. Applied Physics Reviews. 2019;**6**:011303

[28] Kumar SD, Thapliyal H. Design of adiabatic logic-based energy-efficient and reliable PUF for IoT devices. Journal of Emerging Technologies in Computing System. 2020;**16**:34

[29] Cancio M, Yasuhiro T. Low-power two-phase clocking adiabatic PUF circuit. Electronics. 2021;**10**:1258

[30] Cancio M, Yasuhiro T. Ultra-low power FinFETs-based TPCA-PUF circuit for secure IoT devices. Sensors. 2021; **21**(4):8302

[31] Veendrick HJM. Short-circuit dissipation of static CMOS circuitry and its impact on the design of buffer circuits. IEEE Journal of Solid-State Circuits. 1984;**19**(4):468

[32] De V, Ye Y, Keshavarzi A, Narendra S, Kao J, Somasekhar D, et al.

Techniques for leakage power reduction. In: Book of Design of High-Performance Microprocessor Circuits. Wiley-IEEE Press; 2001

[33] Lu Y, Agraval VD. CMOS leakage and glitch minimization for power performance tradeoff. Journal of Low Power Electronics. 2006;**2**(3):378

[34] Panda PR, Shrivastava A, Silpa BVN, Gummidipudi K. Power Efficient System Design. New York: Springer; 2010

[35] Tran DJ, Acuff MJ. Dynamic logic circuit. United States Patent:5,859,547. 1999

[36] Moon Y, Jeong DK. An efficient charge recovery logic circuit. IEEE Journal of Solid-State Circuits. 1996; **31**(4):514-522

[37] Yasuhiro T, Toshikazu S, Yokoyama M. VLSI implementation of a 4x4-bit multiplier in a two-phase drive adiabatic dynamic CMOS logic. IEICE Transactions on Electronics. 2007;**E90-C**(10):2002-2006

[38] Yasuhiro T, Toshikazu S, Yokoyama M. Two-phase clocked CMOS adiabatic logic. Journal of Electronics and Communications. 2009; **3**(1):17-34

[39] Nazrul AN, Yasuhiro T, Toshikazu S. LSI implementation of a low-power 4×4-bit array two-phase clocked adiabatic static CMOS logic multiplier. Microelectronics Journal. 2012;**43**(4): 244-249

[40] Kazunari K, Yasuhiro T, Toshikazu S. Two phase clocked subthreshold adiabatic logic. IEICE Electronic Express. 2015;**12**(20):1-12

[41] Cancio M, Apolinario M, Yasuhiro T. Low power source biased semi-adiabatic logic circuit for IoT devices. In: Proceedings of the International Symposium on Intelligent Signal Processing and Communication Systems (ISPACS'18). Okinawa-Japan; 2018. pp. 43-47

[42] Tiri K, Akmal M, Verbauwhede I. A dynamic and differential CMOS logic with signal independent power consumption to withstand differential power analysis on smart cards. In: Proceedings of European Conf. Solid-State Circuits (ESSCIRC '02). Firenze, Italy; 2002. pp. 403-406

[43] Bucci M, Giancane L, Luzzi R, Trifiletti V. Three-phase dual-rail pre-charge logic. In: Proceedings of CHES'06, Yokohama Japan; 2006. pp. 232-241

[44] King TJ. FinFETs for nanoscale CMOS digital integrated circuits. In: Proceedings of IEEE/ACM International Conference on Computer-Aided Design. San Jose; 2005. pp. 207-210

Chapter 7

Resistive Switching and Hysteresis Phenomena at Nanoscale

Vithaldas Raja and Ramesh Mohan Thamankar

Abstract

Resistive switching at the nanoscale is at the heart of the memristor devices technology. These switching devices have emerged as alternative candidates for the existing memory and data storage technologies. Memristors are also considered to be the fourth pillar of classical electronics; extensive research has been carried out for over three decades to understand the physical processes in these devices. Due to their robust characteristics, resistive switching memory devices have been proposed for neuromorphic computation, in-memory computation, and on-chip data storage. In this chapter, the effects of various external stimuli on the characteristics of resistive switching devices are comprehensively reviewed. The emphasis will be given on 2-dimensional (2D) materials, which are exciting systems owing to superior electrical characteristics combined with their high stability at room temperature. These atomically thin 2D materials possess unique electrical, optical and mechanical properties in a broad spectrum, and open the opportunity for developing novel and more efficient electronic devices. Additionally, resistive switching due to light has also grabbed the attention of optoelectronic engineers and scientists for the advancement of optical switches and photo tuned memristors. The variety of material systems used in the fabrication of memristors is comprehensively discussed.

Keywords: resistive switching, nanoscale, neuromorphic devices, hysteresis, computation, low dimensional materials, h-BN, MoS_2, graphene, Memristor, neuromorphic computation

1. Introduction

Continuous miniaturization of electronic devices has forced researchers to find newer ways of designing novel materials for nanoscale devices. Among the various devices needed in the electronic circuitry, memory devices represent an important category, with an increasing market. These components save the information in terms of "0" and "1". These two states are attributed to the low resistance and high resistance states of the memristive devices. Resistive switching memory devices represent the simplest group among various types of memory devices. In the simplest design of a resistive switching device, an active material (normally an oxide or a large band gap material in general) is sandwiched between two metal electrodes to form a Metal–Insulator–Metal (M-I-M) junction. When an external stimulus like an electric field is applied across the two metal electrodes, electronic conduction kicks-in at a voltage specific to the active material. This trilayer system without the external stimulus will be in the high resistance state and can be switched using very low applied voltages. As the applied field strength reaches the breakdown strength of the active material, an insulator-to-metal transition takes place, inducing a low resistance state. The transition is characterized by a rapid increase in the current of 2–5 orders of magnitude depending on the material system considered. This typically happens because of the formation of conducting filaments parallel to the current flow, which is either due to a local phase transition, Joule heating or field-induced carrier generation. When these devices reach a nanometer size scale, electrons can tunnel through the active material (tunnel barrier), giving rise to a leakage current. To overcome this issue, researchers have been trying to replace the existing oxide materials (Si-based oxides) by newly designed materials with higher dielectric constant.

Resistive switching (RS) phenomena are described using various characteristics that define performance and device stability. These characteristics include volatile/non-volatile type, operating voltage, switching speed, ON/OFF ratio, endurance, and retention. The RS process is also classified as unipolar or bipolar based on the polarity of the operating voltage. Various intrinsic and extrinsic factors play an important role in the resistive switching characteristics. Based on the requirements of future technologies, ideal values of important characteristics should be the following: i) operating voltages <1 V, ii) power consumption $\sim$10 pJ per transition, iii) switching time or transition speed from one state to another below 10 ns, to name a few. A stable resistive switching device is expected to have endurance greater than 10^9 cycles, with data retention for more than 10 years. For higher storage density, the cell size of a unit (M-I-M) structure should be 576 nm^2 with an ON/OFF ratio of 10^6. However, the main constraint for these devices to be ready for applications is the identification of a single system with characteristic values equal to or near to these ideal values [1].

The hysteresis in the resistance of the switching device has been studied over a few decades now. The hysteresis behavior in resistance was first proposed by Leon Chua who coined the term “memristor”, which is considered to be the fourth element in the classical electronics [2, 3]. It took almost two decades to establish the memristive effect experimentally [4]. The coupled solid-state electronic and ionic transport under the external stimuli was measured and established as a memristor based on $Pt/TiO_2/Pt$ device. Meanwhile, there are enough studies on devices showing the resistive switching behavior. For example, resistance switching based on ion migration [5], organic materials based non-volatile memory elements [6] and solid-state electrolyte [7] and thin film based resistive switching devices [8].

In general, the resistance switching of the device is characterized by an increase in the current level of 2–5 orders of magnitude. The conducting filament created during this transition may or may not be stable at room temperature. If the room temperature thermal energy is sufficient to disrupt this conducting filament, the low resistance state is not stable. In this scenario, a consistent memory performance will not be attained. While, if the conducting filament created during switching is stable at room temperature, the low resistance state will be stable at room temperature. Typically, the metal oxide-based memory devices are expected to be stable beyond 80°C for a given duration. When cyclic external stimulus is applied to such a device, a repeatable switching between high resistance state (HRS) to low resistance state (LRS) would result in hysteric current – voltage characteristics.

For the usage of resistive switching devices in information processing architectures, a high resistance state (HRS or OFF state) and a low resistance state (LRS or ON state) should be clearly defined. This should be followed by fast switching. To achieve this goal, extensive research has been carried out to optimize the design of the device structure, the materials used for the devices and size scale of the devices. Over the past two decades, since the discovery of graphene, many 2D materials have been also proposed/used for the fabrication of resistive switching devices.

The present chapter is divided into three main sections. Section 1 gives an introduction to the resistive switching process and its characteristics. Section 2, discusses the hysteresis in the electric field-induced resistive switching devices and different material systems used in the M-I-M structures. In Section 3, light-induced resistive switching is described. Optically-induced resistive switching will enhance the usage of such M-I-M devices in the field of optoelectronics and optical sensing. The last section, Section 4, is focused on the resistive switching devices used in neuromorphic devices and computation. The action of synapse – neuron pair in brain-inspired information processing is mimicked in the laboratory by fabricating resistive switching memory devices and architecture. At the end, Section 5 gives the brief conclusion of the chapter highlighting some of the objectives to be achieved in future nano-electronic technologies.

2. Electric field-induced hysteresis in resistive switching devices

2.1 Resistive switching mechanisms

One of the common external stimuli used in electronic devices is the electric field. **Figure 1** (Adapted from [9]) shows an overview of how a typical resistive switching device and a combination of devices can be used as a memory architecture. **Figure 1(a)** shows a schematic diagram of a metal – insulator – metal (M-I-M) device. These individual memory devices can also be integrated into an architecture to process information, as shown in **Figure 1(b)**. Here, the resistance of the individual memory element can be set at desired values by applying voltage pulses. Extensive research has been carried out to understand the underlying mechanism of hysteretic resistive switching in these simple devices. Two main proposals to explain the resistive switching process are represented by: i) the formation of conduction path due to metallic filament as shown in **Figure 1(c)**. This is generally referred to as conducting bridge memory (CB) or electrochemical memory (ECM) or even sometimes referred to as programmable metallization (PM) device. ii) the presence of oxygen ion vacancies, which will form a conduction path due to the applied electric field, as shown in **Figure 1(d)**. In this case, the migration of oxygen ion vacancies is considered to be the main reason for the low resistance state.

In the first model shown in **Figure 1(c)**, the formation of a metallic filament across the active material causing a short path for electron transport is considered. This is termed conductive–bridge random access memory (CBRAM) and there are a lot of reports available in literature to support this argument [10, 11]. This phenomenon is significantly affected by the type of electrode material used in the devices. In the second model, **Figure 1(d)**, charged oxygen ion vacancies form a conducting bridge between the top and bottom electrodes of the device, thus resulting in a switching. This model is based on ions migration, where field-driven migration of oxygen ion vacancies or metallic ions occurs through the active material. Both processes have been explained using analytical and numerical models [12, 13]. The analytical model explains the phenomena occurring at the device level considering the circuit design

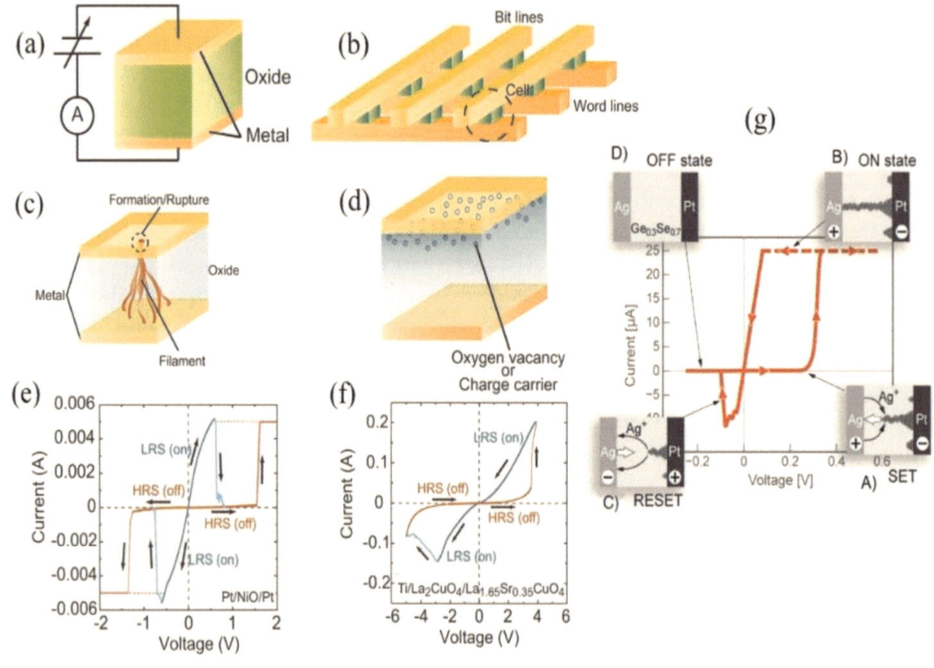

Figure 1.
(a) Schematic diagram of a resistive switching device (M-I-M) consisting of two metallic electrodes separated by the switching material [9] (b) Combination of the devices in the memory architecture. (c) A typical metallic filament formed between the two electrodes giving a low resistance state. (d) The process-induced oxygen vacancies randomly distributed in the oxide material is depicted. These oxygen vacancies form a chain when a suitable voltage is applied to the device and the chain will act as a conduction path. Typical current – Voltage characteristics showing the hysteresis behavior depicting the switching process are shown in (e) and (f). The unipolar switching is shown in (e). Here, a forming process, rupture of the filament during the RESET process and filament formation during the SET process is shown. In (f) a bipolar switching is shown [9]. In this case, the electrochemical migration of oxygen ions is responsible for the hysteric current–voltage behavior of the device. (g) Typical current – Voltage characteristics of Ag-AgGeSe-Ag switching device. The insets a to D depict the various stages of the switching cycle.

parameters. The numerical models are based on the parameters affecting the operation of the device, such as the type and shape of the bias voltage, and compliance current. The compliance current is set to limit the current flowing through the device, such that the device can be operated successively for many cycles. Without current compliance, the applied voltage might create a complete dielectric breakdown and the device might not be able to get back to the high resistance state. As of now, the most widely-accepted model explains the switching process via trap-assisted tunneling mechanism (TAT) [14, 15]. In this case, the tunneling of electrons occurs via the traps generated or recombined with certain probabilities depending on the local electric field. **Figure 1(e)** and **(f)** show the typical hysteretic behavior of resistance switching in two types of devices. When the voltage is ramped across the device, the device switches from a high resistance state (HRS) to a low resistance state (LRS) at a well-defined voltage, with a clear distinction between the two resistance states. The critical voltage at which this transition occurs depends on the thickness and dielectric constant of the active material present in the device. The fact that the current – voltage characteristics show hysteresis implies that these structures can be used as memories devices to store information. The advantage of such devices is that the memory can be erased by applying a voltage in the opposite direction, which switches the device back to its original high resistance state.

Initially, when the applied voltage is zero, the device is in the high resistance state (OFF) and the switching material is in its pristine state (**Figure 1(g)**). When sufficiently large voltage is applied, the device turns ON (SET process). Here, a stable metallic filament grows through the switching layer by means of anodic dissolution of metal, as shown in the inset A of **Figure 1(g)**. Further application of higher voltages will strengthen this metallic filament and a stable ON state (low resistance) is retained. This is due to the migration of the metallic ions generated and the crystallization of the metallic filament. When a sufficiently large bias in the reverse direction is applied, the formed metallic filament dissolves (inset C) resulting in low current passing through the device (HRS), which results in the RESET state of the device. This forming and breaking of the metallic filament is considered as electrochemical metallization and is found to be the main cause of the memristive effect [16]. In general, the electrochemically active metals like Ag and Cu are used to study the electrochemical metallization phenomena in devices.

The Valence Change Mechanism (VCM Type) is attributed to the switching mechanism in devices based on binary oxides or even multinary oxides. In this case, the electrochemical dissolution of the metallic electrode does not occur. As early as in 1960's, studies were carried out on Nb-Nb_2O_5-B and Nb-Nb_2O_5-In trilayer structures [17], and later in 1990's, mostly manganates, titanates and zirconates have been studied. The trilayer structures based on these materials exhibited bipolar switching [8, 18]. Since the electrode metallization does not occur, a SET procedure is always needed before the device shows a bipolar switching. Several mechanisms have been put forward to explain the mechanism of resistive switching in this case. The most favored mechanism is the charge-trap model. Here, the injected charges are trapped at the site of defects within the insulating material of the device [16]. In the case of strongly correlated electron systems, which undergo insulator-to-metal transition (IMT), the charge injection acts as doping, resulting in the switching of the resistance state of the insulator. Further, there are extensive reports on the anion-assisted switching, mostly based on the oxygen ion vacancies, which are more mobile than the metal cations. The formation and dissolution of the conduction path due to oxygen vacancies have been proposed as a reason for the resistive switching. The created anion vacancies will affect the charge state of the metal cations, and thus, this generic phenomenon is normally referred to as valence change memory (VCM).

2.2 Resistive switching characteristics

The ability of resistive switching devices to retain a certain resistance state (HRS or LRS) even after the application of a voltage over a period of time makes them non-volatile. Thus, a very low voltage (lower than the operating voltage) is required to read, leading to the low power consumption of such devices. Instead, for volatile devices, the write and read voltages are of equal weight and the device is brought back to its original state, upon removing the power supply. Thus, for random access memory technology, non-volatility is essential. These non-volatile resistive switching type devices are also proposed for neuromorphic computing architecture, as electronic synaptic devices. For such computing architectures, there are two types of rules to be implemented, short-term plasticity (STP) and long-term plasticity (LTP).

STP can be achieved either by using volatile devices or complex architectures based on non-volatile devices [2]. This has encouraged several research groups to develop resistive switching devices showcasing volatile behavior. Shi et al. [19] were the first to demonstrate the coexistence of volatile and non-volatile behavior in resistive switching devices based on hexagonal-Boron Nitride (h-BN). By means of modulating the compliance current, the coexistence of volatile and non-volatile characteristics is studied. It was also due to the inherent chemical stability of the material that it exhibited a controlled dielectric breakdown. The h-BN device was able to completely recover the resistance state upon removal of the voltage supply. The hysteresis curves for h-BN devices are shown in **Figure 2**. A schematic diagram of the Au/Ti/h-BN/Cu switching device is shown in **Figure 2(a)**. Resistive switching curves indicate volatile behavior as shown in **Figure 2(b)**. The volatile behavior can be seen on both sides of the voltage cycle. The device regains its high resistance state (HRS) when the voltage is cycled via 0 V - V_{max} – 0 V. Because of the chemical and thermal stability of h-BN, the soft breakdown occurring can be completely recovered once the electrical bias is removed. The complete recovery of the high resistance state can be achieved on both bias polarities. **Figure 2(c)** and **(d)** show the volatile switching characteristics with the compliance current values of 1 μA and 10 μA respectively. Even though large scatter is observed in the switching voltages in the lower compliance regime, the hysteresis can be clearly seen, indicating that the device can be used as a memory device. It can also be observed that the voltage range at which the device can be operated as a memory element is also larger when a higher compliance current is used. This can be due to the unstable conductive filaments (CF) being formed when the device is operated with a lower current compliance value compared to more stable CF, and thus wider hysteresis, when higher current compliance is used [9]. Further, it is interesting to notice that non-volatility can be activated by setting a higher compliance current. This is clearly shown in **Figure 2(e)** and **(f)** for compliance currents of 500 μA and 3 mA, respectively. **Figure 2(e)** shows the non-volatile behavior when a positive SET voltage cycle is applied, and **Figure 2(f)** shows the non-volatile characteristics when a negative SET voltage cycle is applied. Here, also the area of the hysteresis is different for different SET voltage cycles applied. The switching characteristics with negative SET cycle seem to show a sharp transition from HRS to LRS state.

It is important to note that the complete recovery of the high resistance state in the case of volatile switching is shown in the top row. This is unlike in the case of transition metal oxide switching devices. Even though h-BN shows extreme chemical stability, together with a reliable device performance, there are still many "unknown" parameters that must be understood in detail when 2D materials are used in the devices. Some of the 'unknowns' are highlighted in the recent work by Pey et al. [20] as indicated in **Table 1**.

Most resistive switching devices are based on thin films of transition metal oxides (TMO) as active elements in M-I-M structures. A large number of systems exhibiting resistive switching characteristics have been reported in literature. As mentioned earlier, there is a variety of resistive switching characteristics for this simple structure of the device. For a decade or so, there has been a common consensus on the phenomenology of resistive switching in these devices. Switching phenomena can be classified into unipolar and bipolar memory effects. One can term this effect as "switching modes" in the devices. Extensive research in 2D materials has extended the horizon of unipolar and bipolar effects observed in a variety of 2D materials.

Both unipolar and bipolar resistive switching processes have been observed over the years. For unipolar resistive switching devices, a voltage with the same polarity is applied for both SET and RESET process. This means that the memory state of the

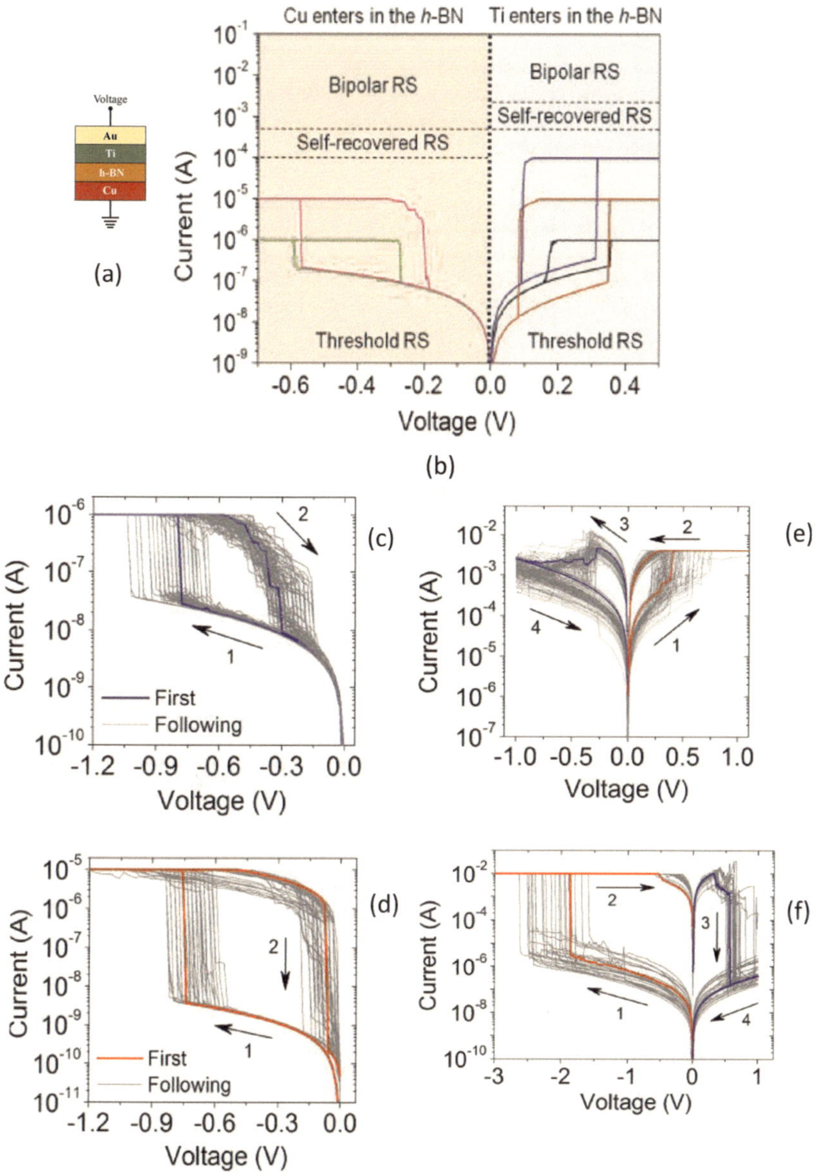

Figure 2.
The two types of resistive switching observed in Au/Ti/h-BN/Cu based resistive switching devices [19]. (a) Schematic diagram of Au/Ti/h-BN/Cu switching device (b) volatile resistive switching characteristics in both polarities. On both sides, the device goes back to its HRS state when the voltage is cycled through a full cycle. (c) the volatile resistive switching of the device with a compliance current set at 1 μA. Clear hysteresis of the current – Voltage characteristics can be observed. (d) Similar volatile resistive switching characteristics of the device with compliance current of 10 μA. Higher compliance current setting seems to stabilize the hysteresis with smaller deviation in switching voltage compared to the lower compliance. (e) the non-volatile switching characteristics of the same device. In this case, a negative voltage cycle is applied as a SET cycle for the device. (f) the nonvolatile switching characteristics with positive SET cycle. In both cases of positive and negative SET cycle applied, a clear switching characteristic is achieved.

device can be switched by successive application of biases with same or opposite polarity. This is because the devices are primarily governed by thermochemical mechanisms, such as Joule heating effects in the formation and rupture of conductive

Unknowns in breakdown phenomena in 2D materials [20]
1. Type and charge state of vacancy defects causing breakdown.
2. Validity of percolation model for the area and thickness scaling
3. Origin of defect clustering – intrinsic or extrinsic or both?
4. What are the activation and relaxation energies of the defects?
5. Role and competition between charge trapping and SILC in h-BN.
6. Size and shape of percolation path
7. Role of metal – h-BN interface on the breakdown kinetics
8. Role of self-heating and its impact on the post-breakdown behavior
9. Role of covalent B-N bonds and Van der Waal forces on percolation
10. Which extrapolation model could be used for time dependent breakdown lifetime estimation?
11. Overall conduction path propagation in layered materials

Table 1.
List of issues which are still to be understood in detail in the case of 2-D materials, as listed in ref. 20. The 'unknowns' listed in this table will be a key research topic for the next decade when large band gap 2-D materials will be considered.

filaments. Meanwhile, the bipolar resistive switching process is associated with electrochemical mechanisms, which require voltage of opposite polarity to switch between the memory states. These distinct behaviors of polarity (unipolar/bipolar) and volatility (volatile/non-volatile), allow identifying suitable applications, where the efficiency of the devices can be exploited.

Interestingly, in some specific cases, the mode of resistive switching is interchangeable. In the case of a bilayer stack of Ga_2O_3/Cu_2O grown on ITO substrates such interchangeable switching was observed. The bipolar switching was attributed to the existence of the traps in the interface and the unipolar switching was ascribed to the oxygen ion vacancies accumulating at the interface [21]. Specific systems have been designed to showcase the coexistence of unipolar and bipolar phenomena. Some of these devices exhibit irreversible transitions from one switching mode to the other (unipolar to bipolar), which are of less use compared to the reversible transition between these modes.

Also, the coexistence of two switching modes in a single device, gives more degrees of freedom in controlling the geometry/composition of conductive filaments, and eventually leads to development of multilevel resistive switching memories [21, 22]. Zhao et al. [22] demonstrated a reversible transition between the two switching modes in $Ag/MoS_2/Au$ resistive switching device, as shown in **Figure 3**. In this particular case, the devices were fabricated on a flexible polyethylene terephthalate (PET) substrate, as shown in **Figure 3(b)**. Bipolar (**Figure 3(c)**) and unipolar (**Figure 3(d)**) switching can be clearly seen in the device. Notice the polarity of the applied SET process. In the case of bipolar switching, the SET process is carried out by applying a positive voltage, and the switching can be observed on both sides of the applied bias. In the case of unipolar switching, the SET voltage cycle and the switching cycle is observed in the negative polarity. Further, these two modes of switching can be tuned by changing the voltage polarity (see **Figure 3(e)**). The bipolar resistance state mode was observed by the application of +3 V for the SET process in the positive direction. The unipolar-resistance state mode was observed by supplying a negative write voltage of -3 V, as shown in **Figure 3(g)** and **(h)**. In this case, the bipolar state was due to the migration of the silver ions (from the electrode) through MoS_2. Meanwhile, in the case of the unipolar state, the conduction paths through the sulfur ions (S) were

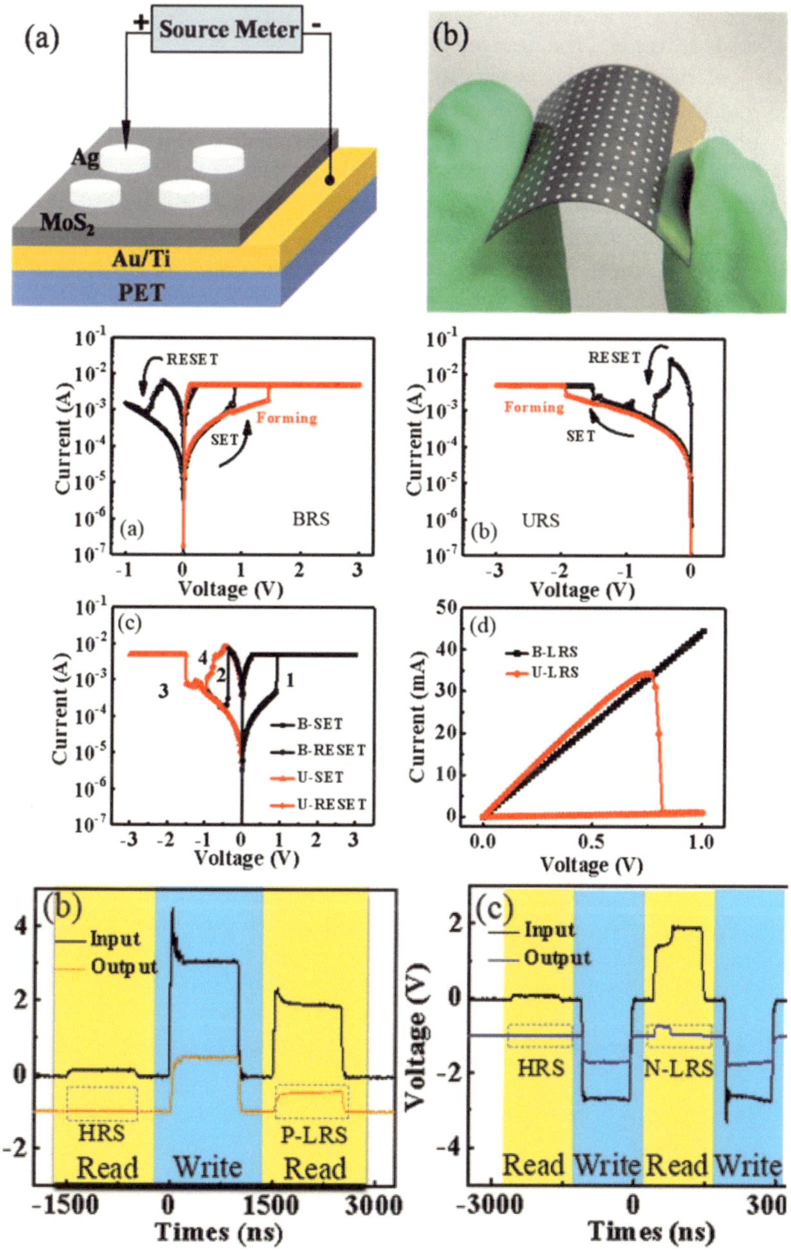

Figure 3.
Flexible resistive switching device. (a) Schematic diagram of the Ag/MoS_2/Au resistive switching device [22]. (b) The flexible PET substrate on which the Ag/MoS_2/Au switching device is fabricated. (c) Bipolar switching characteristics (black) of MoS_2 device with the positive forming voltage cycle (red). Voltage at which the transition to low resistance state (indicated by SET) is clearly defined. A gradual change into high resistance state can be seen in the negative bias region (RESET). (d) a unipolar switching characteristic of MoS_2-based devices. Here, the forming voltage cycle in the switching cycle is in the same polarity of the applied bias voltage. The two switching modes can be altered by changing the voltage polarity as shown in (e). The low resistance state of bipolar (black) and unipolar (red) state has a completely different behavior for positive voltages as shown in (f). The multilevel operation of such device is shown in (g) and (h).

causing the resistive switching. The memory capacity was controlled by modulating the compliance current in the device.

Such modulation of the memory capacity of resistive switching devices was also reported for resistive switching devices based on hafnium oxide [23] and assembled black phosphorus quantum dots (BPQD) [24]. The self-assembled BPQD sandwiched between two poly(methyl methacrylate) (PMMA) polymer layers, is used as a switching layer. This heterostructure is sandwiched between aluminum (Al) electrodes. The switching characteristics are shown in **Figure 4** in detail. In particular, in **Figure 4(a)**, one can notice that BPQD shows a wide memory window, undergoing a HRS to LRS transition at 2.9 V. For bias voltages lower than this critical voltage, the space charge limited conduction dominates the carrier transport. The hysteresis and the switching behavior are maintained for different BPQD film thickness, as shown in **Figure 4(b)**. These devices exhibit exquisite resistive switching curves. The switching voltage for these devices decreases with an increase in BPQD film thickness. This device structure also showed an extremely high ON/OFF ratio (in the order of 10^7) compared to MoS_2-, Graphene-, and Black Phosphorus (BP) nanosheets-based

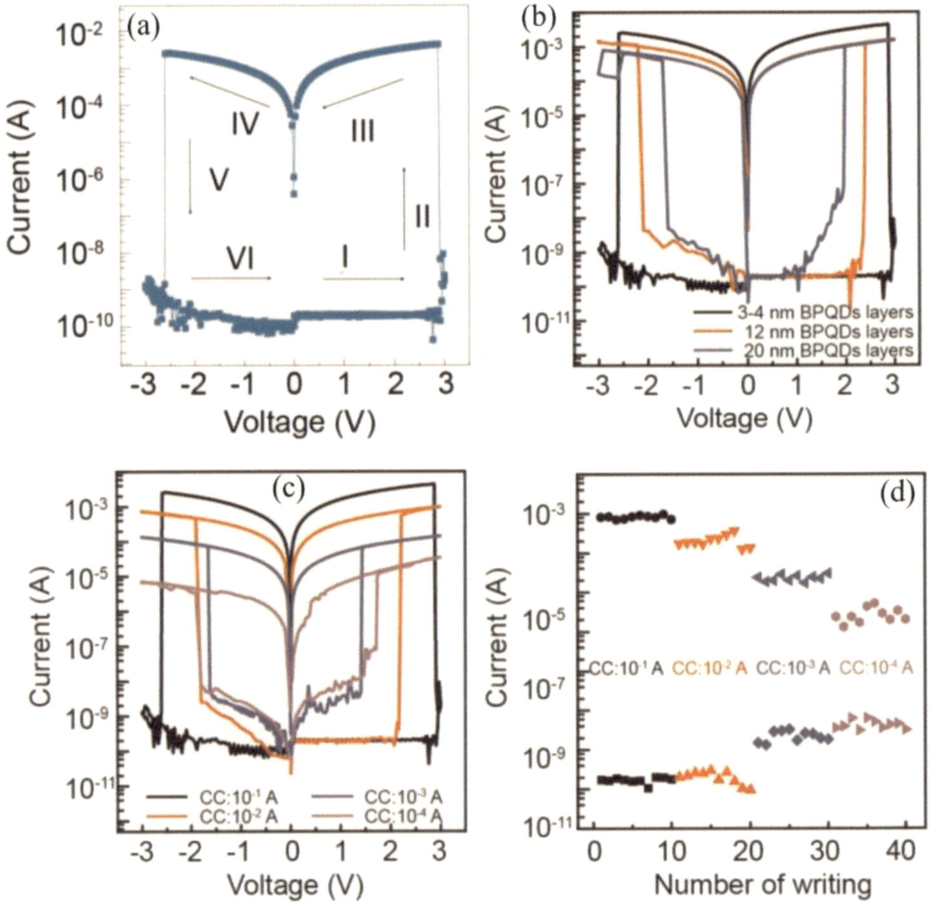

Figure 4.
Resistive switching curves of black phosphorus quantum dot (BPQD) based switching devices [24].
(a) A clear bipolar switching can be seen with bistable resistance state. (b) The ON/OFF ratio can be controlled by modifying the compliance current. (c) The compliance current dependent switching characteristics is shown. (d) The statistics of ON/OFF ratio is seen to depend on the compliance current as shown.

resistive switching devices. For achieving stable multilevel resistive switching, the compliance current in the BPQD device was varied in the range of 10 A^{-1}A to 10 A^{-4}A. This has in turn a direct effect on the LRS and HRS current values and the switching voltages, as noticeable in **Figure 4(c)**. The device also demonstrated tunable resistive switching properties; the ON/OFF ratio depended on the compliance current, as shown in **Figure 4(d)**.

On one hand, materials used for the memory devices form the heart of information processing; however, the device optimization is also an extremely important aspect of nanoelectronics. One of the important parameters in this sense is the switching speed of the memory device. There are reports that illustrate how to determine the switching speed of resistive switching devices. For example, a resistive switching device based on tantalum oxide (Ta_2O_5) switches with an ultrafast speed of around 10 ps at a bias voltage of 3 V [25]. By varying the voltage amplitude, multi-state resistive switching was achieved. The ultrafast performance of this device was demonstrated by using a coplanar waveguide structure of the device with pulsed voltage. The drawback of such device was that the amplitude of the pulsed voltage was extremely high (>3 V). However, at present, the switching speed at reliable voltages (<1 V) is limited by the experimental measurement setup [26].

One of the other important characteristics of resistive switching devices is the operating voltage at which the transition from the high resistance state (HRS) to low resistance state (LRS) takes place. This is expected to be as low as possible, for efficient low power synaptic application in neuromorphic computing architectures. The voltage threshold is dependent on several factors, including the active material between the two metal electrodes, the distance between the two metal electrodes, and the material of the two electrodes. The MoS_2/MoO_x heterostructure can be operated at voltages as low as 0.1 V [27]. The device exhibited symmetric bipolar resistive switching curves with an ON/OFF ratio of $\sim 10^6$. Such devices operating at exceptionally low voltages with a wide hysteresis can be particularly useful for high density storage systems. The resistive switching performance of a device solely based on solution processed non-oxidized MoS_2, displayed linear current - voltage curves in disparity with the heterostructure-based devices. The heterostructure-based device also displayed moderate stability with an endurance of $\sim 10^5$ cycles. Further, VO_2-based nano resistive switches consuming only 4.2 fJ of energy per switching are fabricated by a novel approach [28], whereby VO_2 nanocrystals are embedded in conductive Si tips with an effective lateral size of about 100 nm, after 30mins of synthesis. More importantly, at room temperature, the nanocrystal-based VO_2 devices showed a switching from HRS to LRS at ~ 0.1 V. Additionally, the device showed a high stability of cyclic switching process with endurance longer than 10^{11} cycles. It was clear that such enhanced results were obtained only because of the device geometry, while the VO_2 thin films of ~ 300 nm size grown on silicon substrate had an operating voltage of nearly 6 V at room temperature.

As mentioned earlier, the stability of a resistive switching device is characterized by endurance and retention. Endurance is the measure of the maximum number of stable switching cycles the device can perform between two or more resistive states without progressive breakdown of the device. With the overgrowing demand of modern technologies with dynamic applications, it is necessary for such electronic devices to showcase performance stability over a period of time. Several factors such as (i) structural stability of the resistive switching device, (ii) undesired electrochemical reactions between the electrode and the resistive switching materials, (iii) progressive growth of conductive filament, can lead to higher number of endurance cycles. For a reliable resistive switching device, the ideal endurance value should be >10^9 cycles. There have been few studies in literature that have reported such high endurance values. In a recent publication, Huang et al. [29] reported a bilayer aluminum oxide-based resistive switching devices ($W/AlO_x/Al_2O_3/Pt$) with high endurance cycles from cryogenic to high temperatures (10^7 at 100 K, 10^{10} at 298 K and 10^8 at 400 K). Here, an oxygen-deficient layer of AlO_x was stacked upon a stoichiometric Al_2O_3 layer between tungsten (W) and platinum (Pt) as top and bottom electrodes respectively. This homogeneous bilayer stacking of oxygen poor/rich aluminum oxide layers resulted in controlled activity of oxygen vacancies for all resistive states. Accompanied with these high endurance values, this device exhibited ON/OFF ratio of 10^3 at low switching voltages, with fast switching speed of 28 ns. Engineering of endurance cycles for metal oxide-based RS devices was also projected by Wiefels et al. [30]. They showed that by utilizing ohmic electrodes it is possible to enhance endurance of the resistive switching device by an order of 1–2. Ohmic electrode metals act as a high potential barrier for generation of oxygen defects at metal electrode-switching material interface. This limits excessive formation/accumulation of oxygen defects and uncontrolled filament growth between the top and bottom electrodes.

Data stability is a key aspect for memory storage, synapses of neural network and RRAM applications of resistive switching devices. Data stability is checked by studying the ability of a switching device to retain its discrete resistance states (LRS and HRS) with appreciative ON/OFF ratio after SET and RESET transitions over a period of time. It is essential for the perseverance of the stored data and to avoid fluctuations in the persistent read and write processes of the device. Reliable non-volatile resistive switching devices must have retention value of >10 years at functioning temperature of 85°C. However, this retention value is underachieved by the majority of switching devices, primarily due to the thermodynamical instability of the material and inconsistent formation/rupture of conductive filaments formed during the SET/RESET processes. Nevertheless, a trilayer ($Al_2O_3/HfO_2/Al_2O_3$) oxide-based device showed an excellent reliability with retention of >10 years at a temperature of 85°C [31].

2.3 Effect of external perturbations on RS characteristics

The characteristic results discussed until now are all dependent on either the device structure or the materials used for the active layer and for the electrodes. However, environmental effects also play a role in the switching performance of these devices. Thus, to understand the robustness of resistive switching devices, it is important to test their performance under various extreme conditions. Several reports can be found in the literature till date, where resistive switching of different materials is observed to be influenced by variation of temperature [32], ambient pressure [33], humidity [34] and light illumination [35]. The possibility of remotely controlling a resistive switching device via an external magnetic field has been also reported in a recent study [36]. These external perturbations affect the course of hysteresis of the device, without causing a hindrance to the switching from HRS to LRS. This explicit dependence of the resistive switching characteristics on external stimuli can also pave way for novel integrated technologies. It is also important to note that this dependence on external environmental effects is reversible, i.e., the devices are observed to showcase the original characteristics when returned to normal conditions.

Another material which is extensively studied is HfO_2. The oxides of hafnium and cerium are mostly studied from the point of view of replacing the silicon-based oxides in transistor technology. HfO_2-based resistive switching devices at nanoscale have been extensively studied. The resistive switching as a function of temperature for HfO_2-based resistive switching devices was investigated in a wide range of temperature from 213 to 413 K [32]. A stable bipolar resistive switching was measured at all the temperatures considered. The OFF state current levels of the device increased gradually with increasing temperature. This behavior of the device in OFF state can be attributed to the semiconductor behavior (resistance decreases with increase in temperature). In contrast to that, the device was characterized by a metallic-type behavior in ON state, where the current levels decreased as temperature increased. Layered MoO_x/MoS_2 resistive switching devices showed temperature-dependent resistive switching with well-defined hysteresis in the current – voltage characteristics [27]. The threshold voltage of switching shows an exponential dependence on the voltage sweep rate. Here, the MoO_x is oxygen-deficient and anionic motion results in the resistance switching. Resistive switching characteristics of MoO_x-based devices were studied in the temperature range from 300 to 343 K with jumps of 10 K. Interestingly, both the HfO_2- and MoO_x-based devices showcased similar tuning of the resistive switching with respect to temperature variations. The SET and RESET voltages are observed to decrease with increasing temperature along with an increase in the OFF-state current levels. The hysteresis gradually narrows with increasing temperature, limiting the operating temperature range of such devices.

Importantly, these devices still exhibited stable bipolar non-volatile resistive switching at elevated temperatures. Researchers have been trying to understand the resistive switching devices with increasingly smaller sizes, ultimately reaching atomic scale. For example, an interesting work on atomic scale switching was recently reported in a device based on a solid polymer electrolyte (SPE) comprising a blend of polyethylene oxide (PEO) and $AgClO_4$ [33]. The atomic switching, operating even at elevated temperature, originates from the mobile Ag ions. The report indicated that the operating voltage reduces with temperature and is independent of external conditions, such as air or vacuum. These atomic scale devices are based on electrochemical reactions and ion transport in the electrolyte. Fascinatingly, the devices exhibit reverse reliance when varying temperature in vacuum and air, i.e., the effect of temperature on resistive switching of these devices was inconsistent under air and vacuum. The magnitude of SET and RESET voltages reduced with temperature in the device containing SPE, while the PEO-based device showed an increase in the SET and RESET voltages with temperature. Also, the hysteresis curves appear smooth for measurements performed in vacuum, compared to the abrupt SET and RESET process of the devices tested in air. Additionally, in these devices the change in width of the hysteresis was quite distinct, as shown in **Figure 5(a)**. The hysteresis of the red curve obtained at lower temperature (i.e., −40°C) is quite large. As the temperature is increased, a reduction in the hysteresis can be seen, even though a clear resistance switching is observed. In this low temperature regime, the threshold voltage at which resistance switching occurs is higher than 1 V. By increasing temperature from room temperature, the threshold voltage drastically reduced (**Figure 5(b)**). As indicated in the schematic inset of **Figure 5(b)**, the switching behavior is attributed to the Ag ions dissolving in the SPE matrix, thus forming a conduction path leading to a low resistance state. As the temperature is increased up to 60°C, a clear resistance switching can be seen, with a reduction of the hysteresis width with increasing temperature. A similar temperature-dependent hysteresis can be seen in a simple, lateral metal-VO_2-metal device, as shown in **Figure 5(d)**. The switching voltage reduced drastically with temperature and still shows a sharp transition into the low resistance state [37]. This reduction in the threshold voltage can be seen when the device is operated in ambient

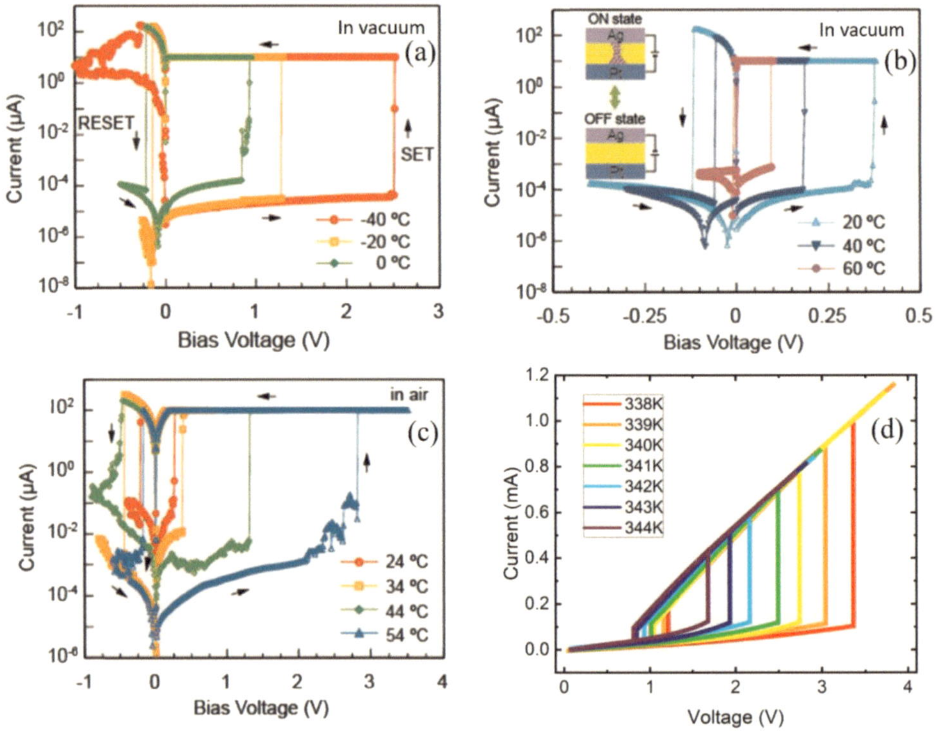

Figure 5.
Resistance switching in Ag/solid polymer electrolyte (SPE)/Pt -based atomic switches [33]. (a) A clear hysteric switching is observed at low temperatures with threshold voltages in the range 1 V–3 V for the temperature indicated. (b) The resistance switching at higher temperature with drastic reduction in the threshold voltages. Both (a) and (b) are measured in vacuum, while the switching characteristics depicted in (c) are measured in ambient conditions. Again, the threshold voltages are higher than 1 V, while the area of the hysteresis reduces with increasing temperature. (d) A similar temperature dependent resistance switching seen in VO_2based device. The area of hysteresis reduces as the temperature is increased. The shape of the switching characteristics is typical for electric field-induced insulator-to-metal transition in VO_2-based devices [37].

conditions. Here, the threshold voltage is preceded by a gradual change in the resistance, differently from the case of vacuum measurements. Similar temperature dependence is also seen in the case of a Cu/Ta_2O_5/Pt atomic switch [38].

Inspired by these temperature-dependent studies, humidity-dependent resistive switching studies for multilayer VO_2 devices [34], at room temperature, have been reported recently. Humidity is present everywhere and strongly affects the performance of electronic devices, especially at the nanoscale. In the VO_2-based switching devices the operating voltage is exponentially reduced with increasing humidity. An increase in adsorption of water molecules at higher humidity levels, eventually assisting the material by increasing the conductivity, required relatively low voltage to switch from HRS to LRS. Once again, the devices exhibited bipolar resistive switching at all humidity levels (11–90% RH) with a decrease in the width of the hysteresis curve at ascending relative humidity. Such strong dependence of the resistive switching devices on relative humidity can be exploited in developing novel robust humidity sensors.

The dependence of resistive switching on the relative humidity levels has been observed in other ECM devices containing $SrTiO_3$, CeO_2, TiO_2, $BaTiO_3$, SnO_2, ZrO_2: Y_2O_3, especially when the oxide material is deposited using physical vapor deposition techniques [39–43]. This process will result in a porous oxide switching layer, which will enable the incorporation of hydroxyl ions. This is detrimental if the oxide material is used in transistor designs where a gate dielectric (an oxide material) is used. Due to the adsorption of water molecules, the operating parameters will be affected. The temperature variation in the operating conditions can affect the filament formation (SET process) and the dissolution of the filament (RESET process) [44, 45]. Various mechanisms have been proposed for oxidation of the copper electrodes used in the $Cu/Ta_2O_5/Pt$ structures [44]. The device characteristics depend on how the Cu ion migrates in the Ta_2O_5 material. Several rate limiting mechanisms have been proposed. Firstly, the copper can get ionized at the oxide interface due to the reduction of the tantalum oxide or copper can get ionized via interaction with ambient gasses like O_2. Further exposure to H_2O molecules can result in the formation of CuO, Cu_2O and also Cu_3O_2 at the oxide interface. These oxides of copper will form the sources of metal ions, which will then migrate through the oxide layer affecting the resistance state of the device. It should be further noted that during the initial stages adsorption of water molecules on VO_2 and Ta_2O_5 surface results in a hydrogen bond network.

Another external stimulus used for manipulating the resistance states of a memristor device is optical light. First of its kind, a graphene oxide-based optical memristor device showed a repeatable resistive switching under ultraviolet (UV) light [35]. Two kinds of devices, lateral and transverse, were developed using graphene oxide (GO). The ON/OFF ratio was found to increase with the graphene oxide thickness. In the transverse type device, GO sheets with effective thickness of $\sim$500 nm was sandwiched between Indium Tin Oxide (ITO) and thermally deposited Ag electrodes; in the lateral devices, the GO layer was drop-casted on interdigitated ITO electrodes with a large effective area. These devices exhibited solely reversible characteristic changes under illumination of long wavelengths of light, and both reversible and irreversible changes in the properties under shorter wavelength. Although narrow bipolar resistive switching was observed in the lateral devices, there was an increase in the current value when the device was illuminated by UV light. The transverse devices offered a wider hysteresis window compared to the lateral devices. However, the hysteresis narrowed under the illumination of UV light and returned to its original state in dark mode. The reversible changes were attributed to photo conductance, whereas irreversible effects are due to the reduction of GO sheets.

Another exciting report was focused on a remotely engineered TiO_2-based resistive switching device using an external magnetic field [36]. The $Ag/TiO_2/FTO$ device showed stable, repeatable resistive switching characteristics, as shown in **Figure 6(b)**. The surprising effect is that the switching process can be controlled by using an external magnetic field. These effects were attributed to residual Lorentz forces, enabling the remote control of resistance states. An exponentially increasing relationship was defined between the SET voltage and the external magnetic field, as seen in the inset of **Figure 6(c)**. The device did not show any degradation, even after repeated resistive switching cycles in presence of magnetic field. It showed a multilevel character under the application of different voltage pulses of 50 μs width under a magnetic field in the range of 0 –2300 Oe, as shown in **Figure 6(d)**.

3. Optically-induced resistive switching devices

One of the novel ways of making a device switch between stable resistance states is by irradiation of light of suitable wavelength. It has attracted considerable attention

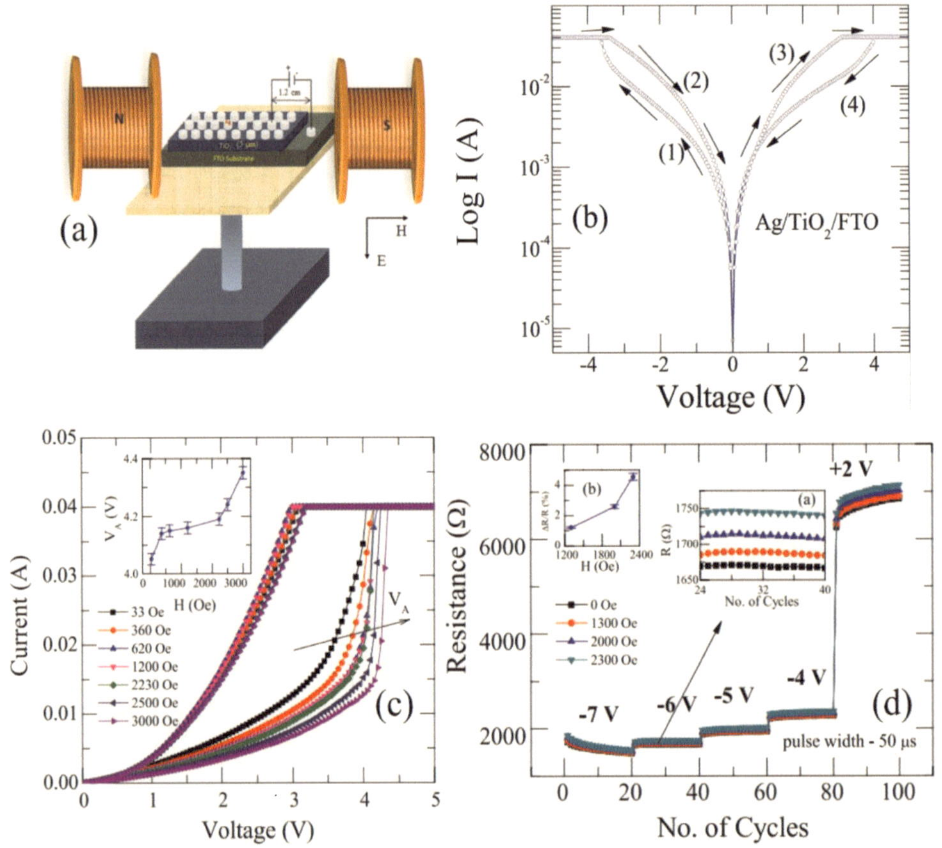

Figure 6.
The magnetic field effect on Ag/TiO$_2$/FTO resistive switching device [36]. (a) A schematic diagram of the experimental set up to study the effect of magnetic field on the resistive switching. The electric field and magnetic field are perpendicular to each other. (b) A typical resistive switching characteristics measured. The device shows a reliable ON and OFF state. (c) The effect of external magnetic field on the switching voltage. The threshold voltage increases with the magnetic field due to residual Lorentz force on the charge carriers. (d) Multilevel resistance states of the device at various voltage pulses indicted.

from the scientific community due to its additional functionalities over the traditional ways of resistive switching. Light of desirable wavelength is capable to induce photo-generated charge carriers in photo-sensitive materials, which can modulate the conductivity of the material. Further, the switching between distinct and stable resistance states or achieving desired programming modes can be controlled by attuning the energy of the photons, i.e., regulating the wavelength of the irradiated light, duration of the illumination, as well as the dark modes and the number of optical pulses for the write and erase process. Compared to the aggressiveness of the electric field-induced resistive switching process, where there is possibility for the devices to undergo an irreversible dielectric breakdown under high electric fields, the optical pulse approach is a much gentler programming procedure and would disregard the possibility of permanently damaging the memristive properties of the device by converting it to an Ohmic material during the switching process. The light-induced resistive switching phenomena would also diminish the occurrence of crosstalk in complex integrated technologies, especially the neuromorphic computing architectures inspired by human brain (**Figure 7**) [46].

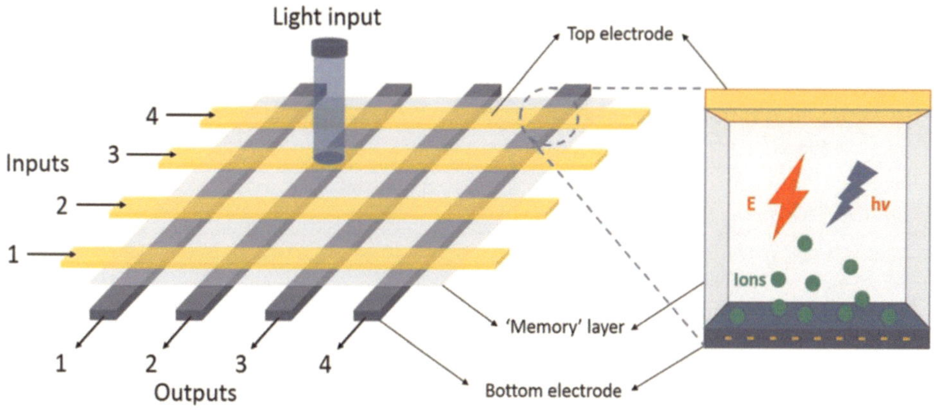

Figure 7.
Optically tunable resistive switching crossbar array. The input and output lines (yellow and blue strips) are separated by the switching material, which is tuned optically. Each trilayer structure is shown on the right. The two metal electrodes are separated by the active material which can absorb light. The absorption of light photons will generate charge carriers inside the material, forming a conductive bridge, thus changing the resistance state of the device. (Reproduced from [46].)

Apart from that, the ability of the optically-induced resistive switching devices to store information when the device switches between stable resistance states with discriminate resistance values, mainly high resistance state (HRS) and the low resistance state (LRS), makes such devices an integral part of the ultra-modern memory technologies. For such photonic memories, the devices are often perceived to encounter a light-induced irreversible resistive switching. This irreversible resistive switching technique can be used as an advantage, permitting the device to function as a "write once read many times (WORM)" type of memory [47]. The irreversible resistive switching would also enhance the encryption of the stored data, as the stored memory/data can be erased solely by the application of suitable external stimulus. These, devices can be possibly switched back to the original HRS by applying suitable biased voltages of appropriate magnitude.

Multifunctionality of a nanodevice is one of the important aspects of the upcoming technologies. For this, hybrid resistive switching devices based on the combined effects of the voltage and optical pulse stimulation have been reported. These devices have displayed enhanced switching characteristics, such as ultra-low SET and RESET voltages, compliance free working devices, and multilevel resistive switching with stable intermediate resistive states. These advanced abilities pave the way for integrated optoelectronic devices, such as light emitting diodes, optical sensors, and photovoltaics.

The selection of wavelength of the optical pulse for inducing resistive switching in a particular material is selected depending upon the bandgap of the material. This is to allow the material to absorb the irradiating photons, which can excite or induce charge carriers in the material. The photogenerated charge carriers assist the material in modulating the active material's conductivity and can be attributed to changes in the resistive switching performance of the device. The choice of the top electrode material can also play a role in these conduction mechanisms, e.g., transparent electrode materials would allow transmission of the photons directly to the active material. Different device structures developed using different oxides, and other 2D materials [47, 48] with corresponding triggering wavelength are tabulated in **Table 2**.

Material/structure	Triggering wavelength/power
ITO/SiO_2/p-Si; ITO-Indium Tin Oxide, SiO_2-Silicon dioxide, p-Si-poly Silicon	VIS/IR light (410 –1100 nm,0.8 μW) [47]
ITO/ZnO/p-Si; ZnO-Zinc Oxide	VIS light (532 nm, 300 mW/cm^2) [47]
Ag/$BiFeO_3$/ZnO/FTO; $BiFeO_3$-Bismuth Ferrite, FTO-Fluorine Doped Tin Oxide	LED (35 W) [47]
Pt/Al_2O_3/SiO_2/Si	UV and IR LED (2.5 mW/cm^2) [47]
QD/GaAs/AlGaAs	CW illumination (2 eV; 730 nW; 44 μW), IR (1.32 eV; 2.2 – 3.6 mW) [47]
Al/PMMA/ZPNPs/PMMA/ITO/QZ	VIS and UV light (0.05 mW/cm^2) [47]
Al/BMThCE/ITO/QZ	UV/VIS light (5.86 mW/cm^2) [47]
Au/HFO/SiO_2Si	VIS light (45 mW) [47]
ITO/HfO_2/ITO	Blue (65 mW/cm) and red (104 mW/cm^2) mediated negative conductivity [47]
Pt/$BaTiO_3$/$NiFe_2O_4$/$BaTiO_3$/Au; $BaTiO_3$-Barium Titanate, $NiFe_2O_4$–Nickle Ferrite	UV (365 nm @ 11.5 mW/cm^2, 302 nm @ 3.78 mW/cm^2) [47]
Au/ZnONRs/FTO/QZ	UV/VIS/IR (200-2500 nm); 300 W Xenon light source [47]
Au/NbNNFs/Au	532/1064 nm @ RT and 405/800 nm @ 8.4 K [49]
PEDOT:PSS/MIM	300/325 nm [47]
Ag/$BiFeO_3$/γ-Fe_2O_3/FTO; γ-Fe_2O_3-Ferric Oxide	White light (20 mW/cm^2) [47]
Pt/AlO_3/SiO_2/Au	UV (300–350 nm; 7 mW/cm^2) [47]
BP/PO_x FET PO_x-Phosphorus Oxide	280/365/660 nm [48]
MoS_2 RRAM	310 nm [48]
Cu/MoS_2NR/Pt	White light (50 W/m^2) [50]
Pd/MoO_x/ITO	365 nm [48]
MoS_2 FET	405–980 nm [48]
h-BN/WSe_2 heterostructure	405/532/655 nm [48]
MoS_2/PbS; PbS-Lead Sulfide	850/1310/1550 nm [48]
MoS_2/hBN/Graphene	458 nm [48]
MoS_2/SWCNT heterostructure	670/820 nm [48]
BP/PZT FeFET; PZT-Lead Zirconate Titanate	808 nm [48]
$ReSe_2$/Graphene	220 nm [51]

Table 2.
A list of 2D materials used in optically tunable resistive switching devices. Column 2 represents the corresponding wavelength or optical power required to induce resistive switching.

Several photosensitive semiconducting materials such as oxides, two-dimensional materials, organic materials, and perovskites have been reported as active materials in optical resistive switching devices [48]. These materials have showed both pure optically-induced resistive switching and reconfigurability of the resistive states by both optical and electrical signals. A simple and straightforward ultraviolet (UV) induced resistive switching was reported for pristine metal oxide switching device treated with PEDOT:PSS [47]. The polymer treatment on the top electrode was known to increase the UV sensitivity of the device. The UV illumination and dark mode treatment cyclically switched the device between steady resistance states. The wavelength-dependent study revealed best results for UV light of 300 nm. The difference between the initial and final resistance state was ~300 Ω. Though the device did not require an additional voltage stimulus for triggering the switching process, a negatively biased voltage signal was needed to bring back the device to the initial resistance state. Another important example of optoelectronic resistive switching devices was demonstrated based upon zinc oxide nanorods [52]. Relying on the desorption of oxygen defects present in the material due to optical illumination, the device showed good switching performance with a photosensitivity of 7.75. The device displays WORM-type memory behavior, retaining the ON state once the illumination is disrupted after forward scanning, as can be observed in **Figure 8(b)**. The switching performance of the device is persistent with optical stimulation, and when the light illumination is cut-off, the switching property also disappears. This behavior is attributed to the absence of photogenerated charge carriers and complete recovery of chemisorbed oxygen vacancies in dark mode.

The diverse electronic and optical properties of 2D materials allow tuning the resistive switching behavior of these devices by a broad spectrum of electromagnetic wavelengths. An important example of white light-induced optical memristor is based on MoS_2 nanorods [50]. A clear bipolar resistive switching with a clearly defined resistance level can be seen in **Figure 9**. The schematic of MoS_2 nanorods based memristor is shown in **Figure 9(a)**. Vertically aligned MoS_2 nanorods are sandwiched between copper and platinum electrodes, where a white light is irradiated on the top electrodes (Copper electrodes) (**Figure 9(a)**). The switching device shows a reduction in the SET and RESET voltage due to the absorption of white light as shown in

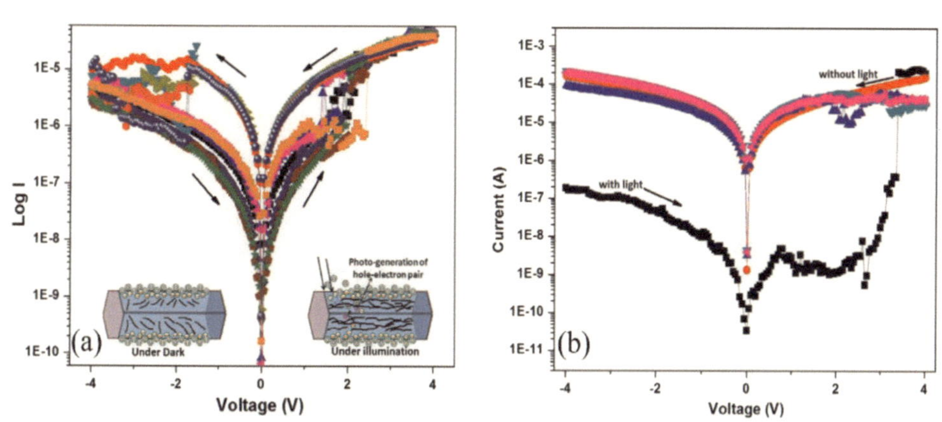

Figure 8.
Light-assisted resistive switching in solution-processed zinc oxide (ZnO) nanorods [52]. (a) Stable bipolar resistive switching is observed under optical illumination of (halogen lamp). Write once read many (WORM) type of behavior shown in (b). The cyclic switching behavior of the device being persistent with the light illumination.

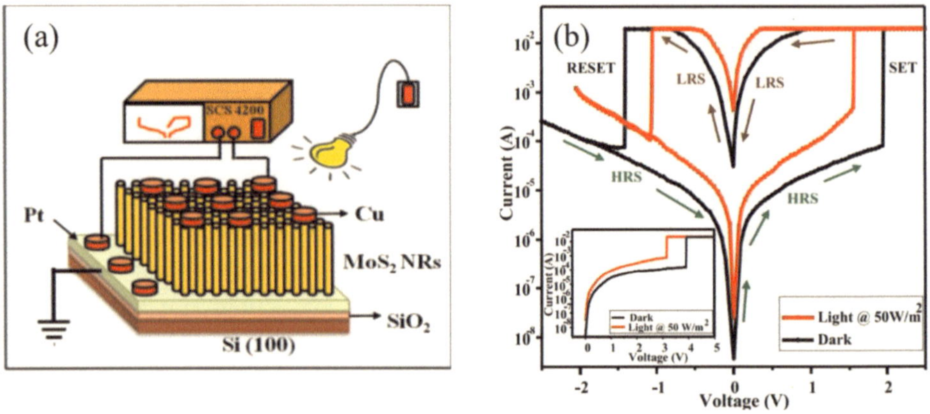

Figure 9.
A typical MoS_2 nanorod memristor device that is switchable using optical light [50] (a) The vertically aligned MoS_2 nanorods are sandwiched between the platinum and copper electrodes. The MoS_2 nanorods undergo resistive switching under the illumination of white light. (b) Resistive switching curves under dark state and white light illumination. The electroforming under the illumination also occurs at lower voltages as shown in the inset of (b). The typical hysteric resistive switching also shows a strong dependence of light illumination.

Figure 9(b). The absorbed white light introduces a conduction path in the MoS_2 via sulfur vacancies. Further, the SET voltage reduces with the intensity of white light. This gives an added advantage to tune the SET voltage at a given light intensity and correspondingly the device can be used for information storage. The ON/OFF ratio of $\sim 10^3$ was also maintained over 1500 cycles, implying an excellent stability of the device.

Interestingly, even though graphene displays a zero-band gap, the resistive switching property of graphene can be controlled with light, from noticeably short ultraviolet to long microwave wavelengths; however, due to its semi-metallic nature, graphene is more suitable for electrode materials. An example of tunable resistive switching device based on Rhenium di-Selenide where monolayer graphene (G) with gold (Au) contacts is used as bottom electrode was recently reported [51]. **Figure 10(a)** shows the switching characteristics with gate voltage. Without any gate voltage, the device undergoes a sharp transition from HRS to LRS at ~ +2.3 V (shown in green curve). When the voltage polarity is switched, the device undergoes a transition from LRS to HRS at around 2 V. This switching is attributed to the formation and rupture of Cu filament. The resistive switching properties of the device are engineered using the gate voltage, which consequently varies the Schottky barrier height at the $ReSe_2$/Graphene junction. This variation in barrier height brings about changes in the ON/OFF ratio and operating voltages of the resistive switching device. Further considerable reduction in these values is observed when a deep UV light of wavelength 220 nm is irradiated on the device, as shown in **Figure 10(a)** and **(b)**.

Interesting results were observed when the superconducting Niobium (Nb) thin films were studied for light-assisted resistive switching [49]. At room temperature, Nb thin films showing metallic behavior under normal conditions, exhibited resistive switching behavior when exposed to light from visible to near infrared wavelengths. Unlike the conventional photosensitive devices, this device switched to high resistance state from the initial resistance state upon being exposed to light, as shown in **Figure 10(c)** and **(d)**. Further, a study of the interaction between superconductivity and photoconductivity was also reported in this work. The light-assisted switching

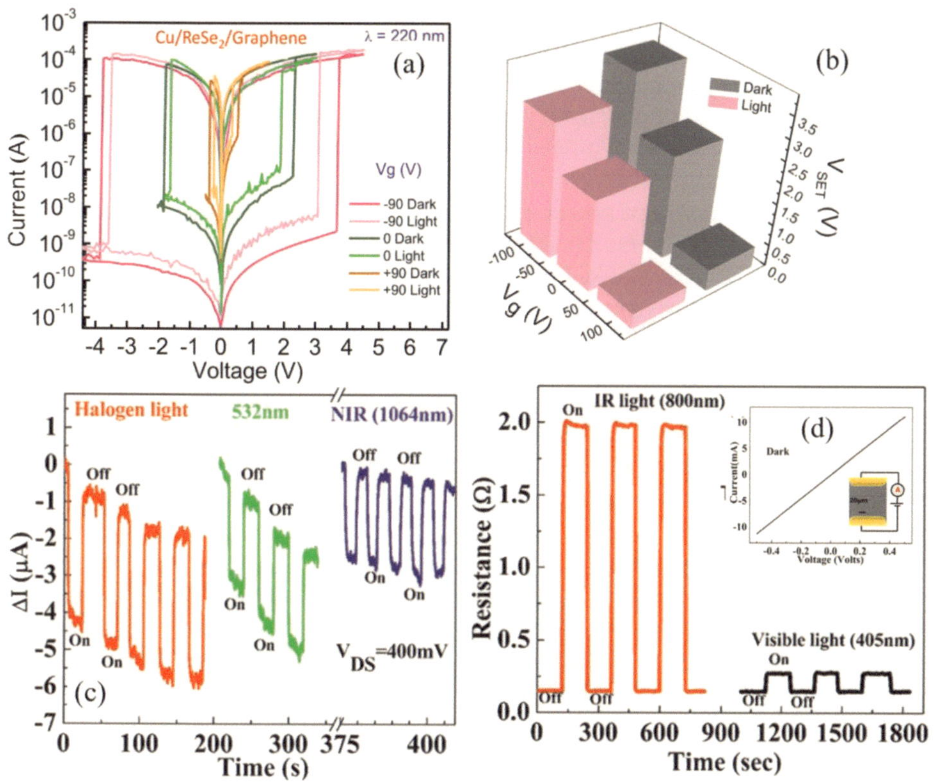

Figure 10.
Light induced resistive switching in Cu/ReSe$_2$/graphene device. (a) Various resistive switching characteristics with gate voltage. Without any gate voltage, switching occurs at ~2.3 V from HRS state to LRS (SET process shown in green). Similarly, when the voltage polarity is changed, the RESET process occurs where the device switches from LR to HRS [51]. (b) Result of the switching voltages in dark and under illumination. (c) the switching characteristics of Nb-based devices under illumination from UV to NIR region. The Nb-based devices goes to HRS under illumination by a halogen lamp, of 532, 1064 nm wavelength, shown in (c) and (d) [49].

behavior of the device at room temperature was explained by photon stimulated electron–phonon scattering. However, at lower temperatures, energy is absorbed by the thin films from the irradiation of light-generated heat in the material. The generated heat increased the effective temperature of Nb thin films, eventually increasing the resistance of thin films. The critical temperature of 8.5 K for the transition of normal metallic Nb thin films to a superconducting state decreases to lower temperatures. The photo-sensing behavior of Nb thin films even in the superconducting state can be exploited for the development of superconducting photodetectors [49, 53, 54].

4. Resistance switching devices for neuromorphic computing

Recently, resistive switching hysteresis has been adopted to mimic the operation of human brain in lab. Human brain can perform multiple operations simultaneously with a fast speed. The brain can process complex and comprehensive information, such as identification, memory, voice analysis and image processing simultaneously with extremely low power consumption (~20 W) [55]. In this way, human brain processes information with much higher efficiency compared to the classical computers based on von-Neuman computation [55–58].

The von-Neuman computation occurs in two different places (central processing unit and memory units) connected by huge number of wires, which becomes a considerable drawback for low power electronics. This necessary requirement of data transfer between the central processing unit (CPU) and the memory units limits the speed and results in increased power consumption in the von-Neuman computing and it is often called as von-Neuman bottleneck [55]. On the other hand, a design of human brain consisting of neurons and synapses offers a better option for fast processing and low power consumption. Efforts have been made to use machine learning software and complementary metal – oxide-based structures have been used to mimic the processing of the human brain. However, this also meets the bottleneck of high-power consumption and the speed of operation, just like the Complementary Metal Oxide Semiconductor (CMOS) architecture in nanoelectronics [56].

Human brain represents a unique architecture where memory and computing take place in the same unit, thereby increasing the data processing ability and reducing the time scale of the information processing. In this way, our brain overcomes the bottleneck of the power consumption compared to the von Neuman. An overview of the brain-inspired neuromorphic device architecture is shown in **Figure 11** - (Adapted from [56]). The independent memory and processing unit connected by physical metallic wires is depicted in **Figure 11(a)**. Since the information storage and processing are performed in separate locations, the processing speed is limited (see **Figure 11(b)**).

The fundamental units of brain computing are neurons and synapses. These are the basis of massive neural networks present inside the human brain. The pre-neuron and post-neuron parts are interconnected via synapses, which offer optimal processing in our brain. The connection strength between two neurons is known as synaptic weight. It is primararily attributed to the volume of neurotransmitter released or absorbed in the biological synaptic action. The action of biological synapses i.e., transmitting of information via electrochemical signal, can be replicated using a two-terminal electronic device, namely a memristor [59]. The two-terminal memristor device can be used to mimic the synapses and neuron combination, as shown in **Figure 11(c)** and **(d)**. Here, the top and bottom electrodes can act as pre and post neurons. The insulating material emulates the biological synapse and the artificial synaptic action is established by modulating the conductance/resistance of the insulating material between these two electrodes. The synaptic actions in neural networks are responsible for advanced activities of the human brain as learning, sensing and remembering. These neural activities are attributed to the synaptic plasticity, which is defined as the synaptic weight modulation ability.

The fundamental rule of synaptic action in the pre-neuron-synapse-post neuron complex is that when the neurons on either side of the synapse are activated simultaneously, then the synaptic weight should increase in the process. This postulate is often referred to as Hebb's rule or sometimes referred to as cell assembly theory [56]. In general, this rule suggests how much the synaptic weight should increase or decrease in proportion to their product when the two neurons are simultaneously activated. The synaptic weight modulation can enhance or depress the connection strength between the two next neurons. This is known as the potentiation or depression of the synapse. For artificial synapse, potentiation can be achieved by positive pulses, and negative pulses can induce depression. Potentiation and depression are the learning and forgetting aspects of artificial neurons [60].

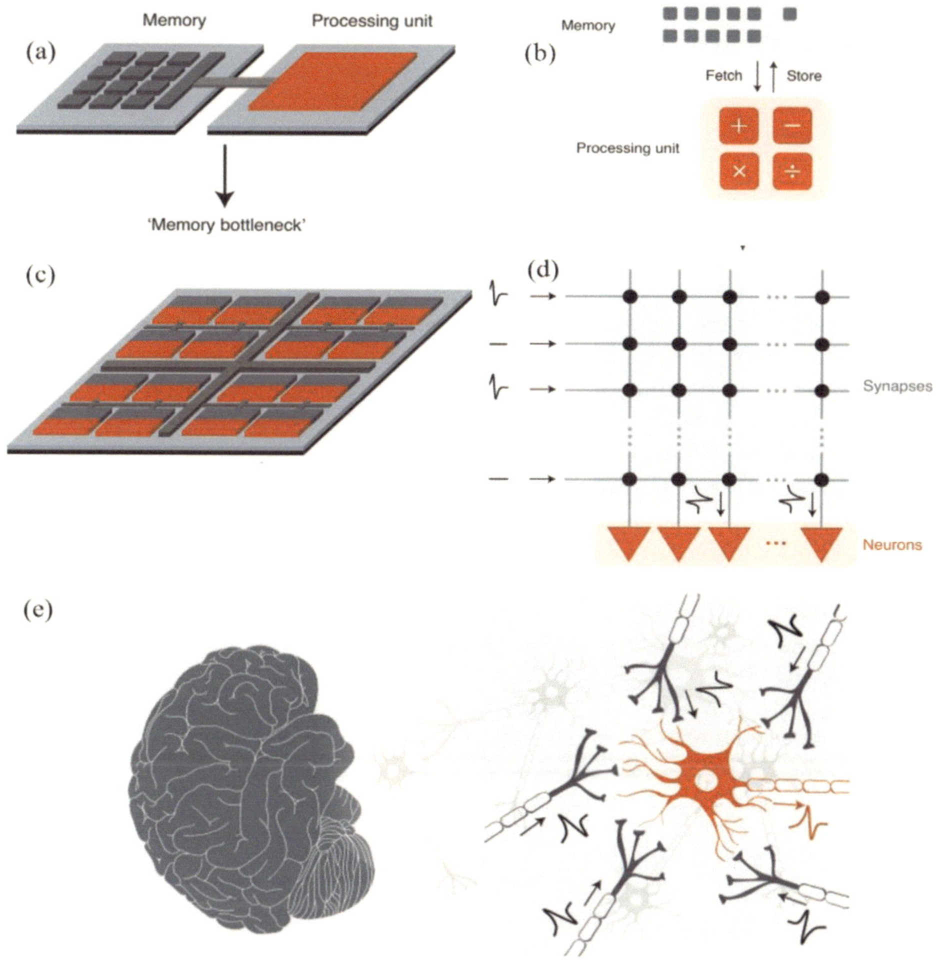

Figure 11.
A schematic diagram of the von-Neuman style of computing and the brain inspired computation [56]. (a) The typical memory and processing unit architecture in the classical computers, which has a bottleneck of processing. (b) A sequential computation in von-Neuman style where data are fetched processed and stored, limiting the processing speed and also the power consumption. (c) A neuron based computational procedure. Multiple input - output connections via synapse – Neuron building block for neuromorphic computation. (d) A typical neuron – Synapse building block of a neuromorphic device. Here, from the multiple inputs via synapses, the neuron processes the information and gives an output indicated schematically. (e) A biological neuron and synapses connections in human brain.

Depending on the modification of synaptic weight, synaptic plasticity is classified into short term plasticity (STP) & long-term plasticity (LTP). When the effects of plasticity last for a short period of time, from milliseconds to few minutes, it is known as STP. For LTP, the retention time of modification effects of the artificial synapse is more than several hours. In the biological brain, STP is accountable for the computational processes like learning, while LTP is responsible for learning and memorizing processes. These properties of STP and LTP can be exploited for artificial neural networks (ANN) using volatile and non-volatile memristive devices [57]. However, by continual stimulation of pulses during the computational processes, one can achieve alteration of synaptic plasticity from STP to LTP. STP is further inspected by paired-pulse facilitation (PPF) and paired-pulse depression (PPD) learning rules. During the synaptic weight modulation process, when two consecutive pre-synaptic spikes stimulate in a short interval of time, the synaptic weight would temporarily increase or decrease. Thus, for PPF the second pre-synaptic spike would produce larger post-synaptic current than the first pre-synaptic spike. Meanwhile, opposite outcome is observed for PPD. Apart from this, synaptic plasticity can also be induced by utilizing the temporal relationship between pre and post synaptic spikes. This phenomenon is known as spike-timing dependent plasticity (STDP) [61].

Two-terminal switching devices include memristors, ferroelectric tunnel junctions (FTJ's) and devices containing phase change materials [62]. In the case of three-terminal devices, these are mainly represented by electrochemical transistors, transistors based on phase change materials and charge trapping transistors [63].

For the design of synapses, the most interesting devices are the memristors, which are trilayer structures with an active material inserted between two metallic electrodes. The dynamic range, multilevel switching and the retention of the memory will determine the quality of the synapse designed. Many materials have been tested for this purpose, as shown in **Table 3**. Although, a variety of active materials have been tested/used in memristive switching devices as indicated in **Table 3**, the research has extensively contributed to the understanding of the fundamental physical mechanisms of the switching process, the switching speed in the devices, as well as designing and fabricating the architecture for the synapse – neuron combination.

The important aspect of these devices is the consistent abrupt transition/switching between high resistance and low resistance states. For an optimal neuromorphic device, a low voltage operation, fast switching and clear distinction between a high resistance state and low resistance state are required. Even though a transistor configuration has been tested for various materials such as $Li_3PO_xSe_x$, $Li_3PO_xSe_x$, Nafion,

Device	Active material
Thin films	Si:Ag $Ge_2Sb_2Te_2$, $Ge_2Sb_2Te_5$, GeSe InGaZnO, AgInSbTe Ta_2O_5:Si, SiO_2:Ni [35] SiO2:Gd [64] MgO, FeO_x, HfO_2, TiO_2, Ta_2O_5, WO_3, NbO_2, Nb_2O_5, ZrO2, Pt/Ag nanodots/HfO2/Pt stacks Ti_3C_2 based MXene Nanosheets Cu^{2+}-Doped $KNbO_3$ ZnO_{1-x}/AlO_y Heterojunction $SrFeO_x$, Nb-Doped $SrTiO_3$, $SrFeO_x$, LSMO
2-D	Graphene, Graphene oxide, MX_2 (M = Mo, W; X = S, Se), h-BN, TMDC, Black Phosphorus, GaSe
1-D	CNT, TiO_2, CuO_x, Cu_2O, NiO, Co_3O_4, Zn_2SnO_4, Ga_2O_3, Ag and Cu nanowires, ZnO nanowires decorated with CeO_2 QDs, PEO-P3HT
0-D	InAs/InGaAs, $CsPbBr_3CsPbBr_3$, Ag,
Polymer based devices	Organic based (PANI), Nanoparticle/Organic, PolyN-vinylcarbazole (PVK), Rotaxene molecule,

Table 3.
Summary of various materials used with hysteretic switching characteristics proposed for neuromorphic computation. The materials are divided into various categories depending on the device fabrication and thickness of the materials used. All the materials are considered from the references listed.

$HfZrO_x$, $Ta_2O_5/HfZrO_x$ and Al NPs/Al_2O_3, the operation speed is fairly large compared to the two-terminal devices.

One common signature of a memristor device is the hysteresis of the current – voltage characteristics. The hysteretic behavior indicates that the device has a long saturation time. The long current saturation time helps to use this device as a neuromorphic device at various frequencies. A device which showed excellent hysteresis is based on Metal - Nb_2O_{5-x} – Metal junctions, displayed in **Figure 12(a)** and **(b)**. The oxygen vacancies are attributed to the conduction via Nb_2O_{5-x}, which are located at 0.2–1.2 eV below the conduction band edge [65]. In this case, the hysteric characteristics are attributed to the long saturation time, which allows a broad range of frequencies for neuromorphic operations that require hysteresis. This device exhibited extremely high capacitance in the low frequency range, where biological process occurs, as shown in **Figure 12(b)** [60]. This is an important feature of the device, which is useful for mimicking the biological process in the laboratory. A similar device performance was seen in a TiO_x nanowire device, as shown in **Figure 12(c)** and **(d)** [60]. These devices could be controlled in the intermediate states of resistance, which is an essential part of the neuromorphic devices. The TiO_2 and Nb_2O_5 based devices represent typical examples of memdiodes to be used in the brain-inspired computation.

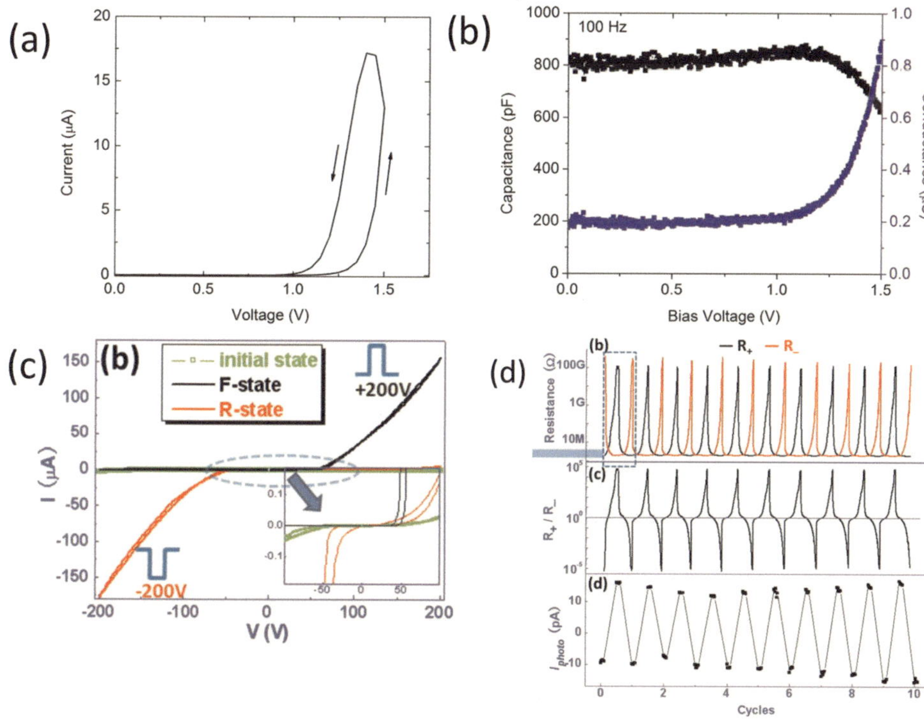

Figure 12.
(a) Typical current – Voltage characteristics of metal – Nb_2O_5 –metal memdiode [65]. Frequency-dependent capacitance indicates extremely high capacitance where the biological processes occur. (b) The capacitance remains almost constant until the threshold voltage is reached. Subsequently, the capacitance decreases, and conductance increases as expected. In the lower row, a device based on TiO_x single nanowire is shown. (c) A small hysteresis can be seen in the low voltage regime indicating a possible operating voltage range for memristors. (d) The endurance performance is shown on the right.

4.1 2D materials for neuromorphic computation

While the heterostructures and epitaxial structures have been studied for their possible application in neuromorphic computations, the low-dimensional materials have attracted attention for this purpose [66].

Materials like graphene, chalcogenides, and hexagonal boron nitride have been tested for their functionality in brain-inspired computation architectures. Recently, neuromorphic systems based on organic polymers have also been tested [67, 68]. Their intrinsic low dimensionality gives an added advantage with respect to their mechanical stability and tunability of the electronic properties. The 2D materials can be integrated into the heterostructure and transistor-type design to get desirable neuromorphic characteristics. Among two-dimensional materials, graphene, and its derivatives with extremely high carrier mobilities are used as electrodes, while systems belonging to the hexagonal boron nitride family with large electronic energy band gap are promising candidates for neuromorphic architectures.

The large band gap of hexagonal boron nitride at the monolayer thickness limit is extremely important for reliable switching characteristics of the heterostructures. Transition metal-chalcogenides are another class of 2D materials that offer wide variety of properties such as semi-metallic and semiconducting behavior. All these 2D materials maintain crystallinity and consequently the electrical characteristics will be more reliable compared to their oxide counterparts in the heterostructures. The existence of crystallinity at low-dimensional scale is essential for faster switching and low energy operations. **Figure 13** displays an overview of the use of 2D materials for neuromorphic computation. Clearly, 2D materials have an advantage over the thin film counterparts since it is possible to intrinsically get materials with a variety of energy gap, as shown in **Figure 13**. With 2D materials, devices can be fabricated in the form of stacked layers, which will give exotic properties and low operating voltages.

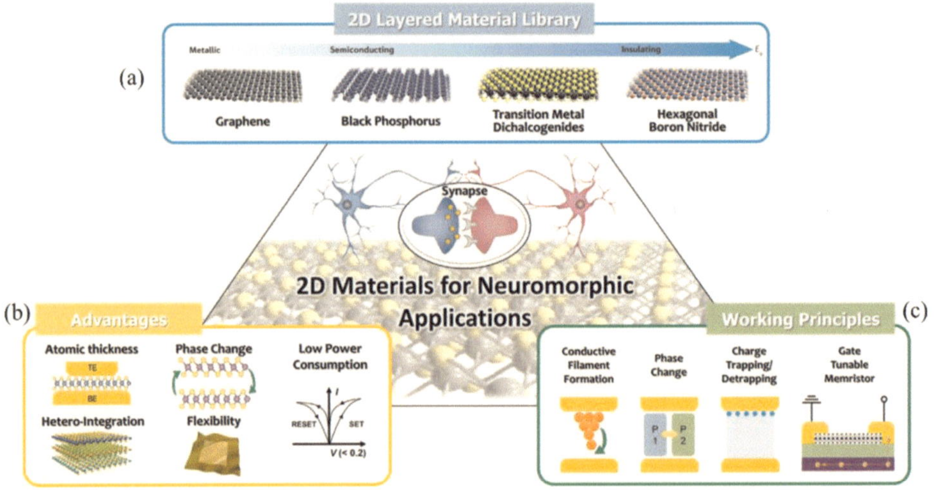

Figure 13.
Overview of the low dimensional materials used for the resistive switching devices [66]. (a) The option of choosing 2-dimensional materials from graphene (no energy gap) to h-BN (~6 eV) gives added advantage in various device structures. (b) Since the materials are in lower dimensions, they will have advantages compared to the bulk materials in terms of the flexibility, low power consumption and most importantly, atomically flat surfaces. (c) Various types of the switching processes attributed to the resistive switching memory devices using the low dimensional materials.

One of the 2D materials commonly used for hysteric resistive switching is hexagonal boron nitride (h-BN), and extensive research is currently being carried out on this material. The h-BN offers a multitude of applications; for example, it can be used as an insulator for a classical transistor design using 2D materials, and a host material for qubits. Having an intrinsic large energy gap, h-BN can form an integral part of the nanoscale 2D transistor design replacing the traditional oxides needed for controlling the ON/OFF states of a transistor (for example, in graphene or MoS_2 based transistors). Further, h-BN also finds importance in the design of quantum bits (QUBITS) for futuristic quantum computation. Nitrogen and Boron vacancies, as well as other defect complexes in h-BN [61, 69–71] have attracted attention due to the possibility of using them as qubits. Recent transmission electron microscopy studies reveal that at the breakdown electric field strength, the electrode metal ions diffuse through the layers creating a conduction path, giving rise to a low resistance state.

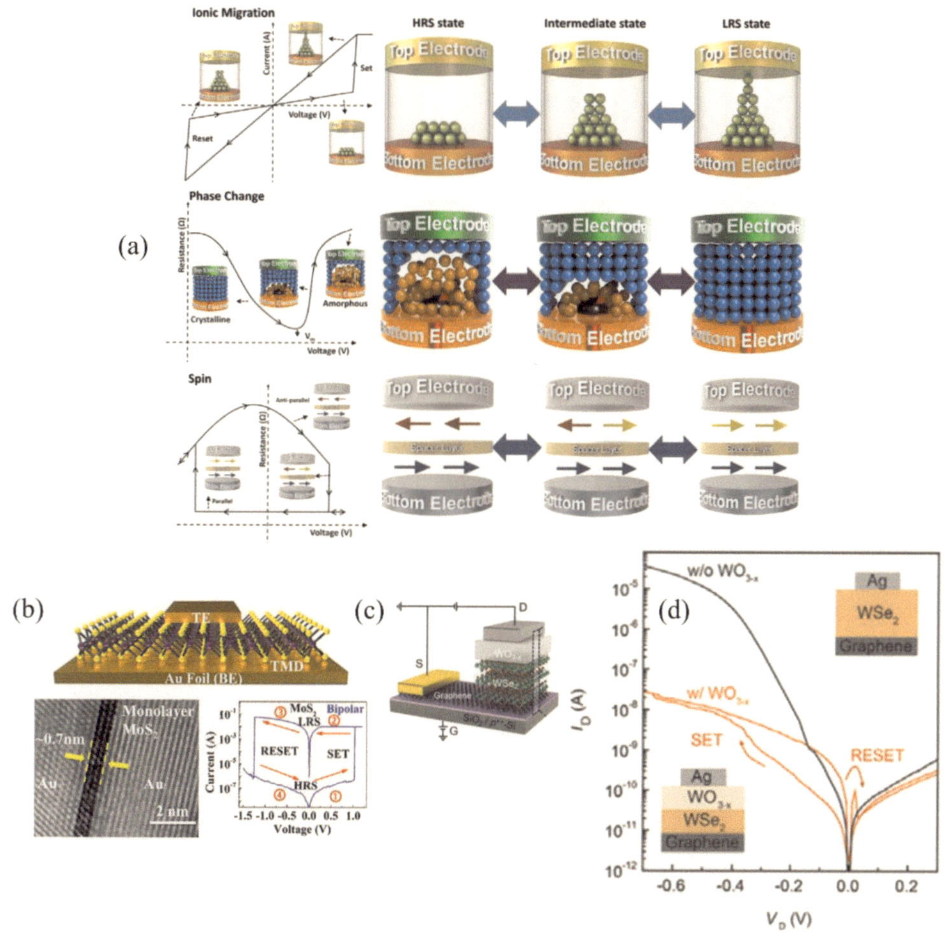

Figure 14.
The mechanism of resistive switching in devices using 2-D materials is schematically shown. The switching can originate from various mechanisms like ionic migration, phase change and the spin flip as indicated as shown in (a). (b) A MoS_2 based neuromorphic device. Transmission Electron micrograph shows clear few layer MoS_2 sandwiched between Au electrodes. This device shows a clear SET-RESET process with well-defined ON and OFF states. (c) A switching device from MoS_2/MoO_3heterostructure. Even though, the hysteresis is small, the device clearly shows a resistive switching as shown in (d).

Recently, MoS_2 double-layer devices have shown spike-timing dependent plasticity (STDP), which is clearly dependent on the hysteretic resistive switching characteristics, as shown in Cu-MoS_2-Au devices [72–74]. On the other hand, the memristor (hysteretic) behavior of monolayer MoS_2/WO_3 heterostructure has also been studied for neuromorphic devices [73]. The resistive switching characteristics of these heterostructures based on 2-dimensional materials can be explained using similar processes observed in the traditional devices, including ion migration and phase change. However, a new phenomenon of spin-flipping is found to be responsible for the resistive switching behavior [71]. **Figure 14(a)** shows the schematic representation of the ion migration, phase change and the spin flipping, which might cause a resistive switching in low dimensional materials and heterostructures. Again, the device configuration is similar to that of the normal trilayer structures containing the active material between the two metal electrodes. An example of such structure, which shows clear resistive switching with clear hysteresis, is represented by the Au/ MoS_2/Au system, shown in **Figure 14(b)**. The layer of MoS_2 is sandwiched between Au electrodes as shown in the transmission electron microscopy images, and a very sharp interface can be seen between Au-MoS_2 on both sides. This device shows a clear bipolar switching as reported in **Figure 14(b)**. Even though the interfaces are separated by atomic distances, the threshold voltage is still about 1 V giving a huge promise for nanoscale, low power memory applications. Similarly, as shown in **Figure 14(c)** and **(d)**, the WSe_2/Graphene heterostructure also shows promising switching properties with hysteresis. Further optimisation of device geometry and operating conditions will yield a better switching performance in these heterostructures.

5. Conclusion

Overall, the future of nanoelectronics and neuromorphic computation revolves around two primary objectives. Firstly, newer materials, heterostructures, and multilayers should be designed to get optimum device performance. The materials used in lab-scale devices should be stable and robust, so that performance would remain acceptable even in the large-scale integration. This will enable not only saving information, but also processing information. The design of materials should address various issues, such as quantum mechanical tunneling of electrons, intrinsic and process-induced defects related to nanoscale and atomic scale devices. These parameters will affect the performance of the devices at the nanoscale. Secondly, the device architecture should be such that these new devices should be compatible to the already optimized device integration and manufacturing. Design aspects should also consider optimizing from the point of view of multiple applications. In this chapter, we have given an overview and discussed materials and devices related to memory applications displaying hysteresis in the current–voltage characteristics. These devices are proposed to be used in low-power nanoscale memory devices and possibly used in the case of neuromorphic computation.

Acknowledgements

Authors thank Vellore Institute of Technology (VIT, VELLORE), India for providing 'VIT SEED GRANT' for carrying out this research work.

Conflict of interest

The authors declare that they have no known competing financial interests or personal relationships that could have appeared to influence the work reported in this book chapter.

Author details

Vithaldas Raja[1] and Ramesh Mohan Thamankar[2]*

1 Department of Physics, School of Advanced Sciences, Vellore Institute of Technology, Vellore, India

2 Centre for Functional Materials, Vellore Institute of Technology, Vellore, India

*Address all correspondence to: rameshm.thamankar@vit.ac.in

References

[1] Lanza M, Wong HS, Pop E, Ielmini D, Strukov D, Regan BC, et al. Recommended methods to study resistive switching devices. Advanced Electronic Materials. 2019;**5**(1):1800143

[2] Chua L. Resistance switching memories are memristors. In: Handbook of Memristor Networks. Cham: Springer; 2019. pp. 197-230

[3] Chua LO. The fourth element. Proceedings of the IEEE. 2012;**100**(6): 1920-1927

[4] Strukov DB, Snider GS, Stewart DR, Williams RS. The missing memristor found. Nature. 2008;**453**(7191):80-83

[5] Waser R, Aono M. Nanoionics-based resistive switching memories. Nanoscience and Technology: A Collection of Reviews from Nature Journals. 2010:158-165

[6] Scott JC, Bozano LD. Nonvolatile memory elements based on organic materials. Advanced Materials. 2007; **19**(11):1452-1463

[7] Kozicki MN, Park M, Mitkova M. Nanoscale memory elements based on solid-state electrolytes. IEEE Transactions on Nanotechnology. 2005; **4**(3):331-338

[8] Beck A et al. Reproducible switching effect in thin oxide films for memory applications. Applied Physics Letters. 2000;**77**(1):139-141

[9] Sawa A. Resistive switching in transition metal oxides. Materials Today. 2008;**11**(6):28-36

[10] Kim KM, Choi BJ, Hwang CS. Localized switching mechanism in resistive switching of atomic-layer-deposited TiO2 thin films. Applied Physics Letters. 2007;**90**(24): 242906

[11] Kinoshita K, Tamura T, Aoki M, Sugiyama Y, Tanaka H. Bias polarity dependent data retention of resistive random-access memory consisting of binary transition metal oxide. Applied Physics Letters. 2006;**89**(10):103509

[12] Ielmini D. Resistive switching memories based on metal oxides: Mechanisms, reliability, and scaling. Semiconductor Science and Technology. 2016;**31**:063002

[13] Ielmini D. Modeling the universal set/reset characteristics of bipolar RRAM by field- and temperature-driven filament growth. IEEE Transactions on Electron Devices. 2011;**58**(12):4309

[14] Grasser T. Tibor, Stochastic charge trapping in oxides: From random telegraph noise to bias temperature instabilities. Microelectronics Reliability. 2012;**52**(1):39

[15] Gehring A, Selberherr S. Modeling of tunneling current and gate dielectric reliability for nonvolatile memory devices. IEEE Transactions on Device and Materials Reliability. 2004;**4**(3):306

[16] Waser R et al. Redox-based resistive switching memories–nanoionic mechanisms, prospects, and challenges. Advanced Materials. 2009;**21**(25–26): 2632-2663

[17] Hiatt WR, Hickmott TW. Bistable switching in niobium oxide diodes. Applied Physics Letters. 1965;**6**(6): 106-108

[18] Asamitsu A et al. Current switching of resistive states in magnetoresistive

manganites. Nature. 1997;**388**(6637): 50-52

[19] Shi Y, Pan C, Chen V, Raghavan N, Pey KL, Puglisi FM, et al. Coexistence of volatile and non-volatile resistive switching in 2D h-BN based electronic synapses. In: 2017 IEEE International Electron Devices Meeting (IEDM). San Francisco, CA: IEEE; 2017. pp. 4-5

[20] Pey KL, Ranjan A, Raghavan N, Shubhakar K, O'Shea SJ. Dielectric breakdown in 2D layered hexagonal boron nitride – The knowns and the unknowns. In: 2019 IEEE International Reliability Physics Symposium (IRPS). Monterey, CA: IEEE; 2019. pp. 1-12

[21] Zhi YS, Li PG, Wang PC, Guo DY, An YH, Wu ZP, et al. Reversible transition between bipolar and unipolar resistive switching in Cu2O/Ga2O3 binary oxide stacked layer. AIP Advances. 2016;**6**(1):015215

[22] Zhao X, Fan Z, Xu H, Wang Z, Xu J, Ma J, et al. Reversible alternation between bipolar and unipolar resistive switching in Ag/MoS2/Au structure for multilevel flexible memory. Journal of Materials Chemistry C. 2018;**6**(27): 7195-7200

[23] Balatti S, Larentis S, Gilmer DC, Ielmini D. Multiple memory states in resistive switching devices through controlled size and orientation of the conductive filament. Advanced Materials. 2013;**25**(10):1474-1478

[24] Han ST, Hu L, Wang X, Zhou Y, Zeng YJ, Ruan S, et al. Black phosphorus quantum dots with tunable memory properties and multilevel resistive switching characteristics. Advanced Science. 2017;**4**(8):1600435

[25] Böttger U, von Witzleben M, Havel V, Fleck K, Rana V, Waser R, et al. Picosecond multilevel resistive switching in tantalum oxide thin films. Scientific Reports. 2020;**10**(1):1-9

[26] Menzel S, von Witzleben M, Havel V, Böttger U. The ultimate switching speed limit of redox-based resistive switching devices. Faraday Discussions. 2019;**19**:197-213, 213

[27] Bessonov AA, Kirikova MN, Petukhov DI, Allen M, Ryhänen T, Bailey MJ. Layered memristive and memcapacitive switches for printable electronics. Nature Materials. 2015; **14**(2):199-204

[28] Prinz VY, Mutilin SV, Yakovkina LV, Gutakovskii AK, Komonov AI. A new approach to the fabrication of VO2 nanoswitches with ultra-low energy consumption. Nanoscale. 2020;**12**(5): 3443-3454

[29] Huang XD, Li Y, Li HY, Xue KH, Wang X, Miao XS. Forming-free, fast, uniform, and high endurance resistive switching from cryogenic to high temperatures in W/AlO x/Al 2 O 3/Pt bilayer memristor. IEEE Electron Device Letters. 2020;**41**(4):549-552

[30] Wiefels S, Von Witzleben M, Hüttemann M, Böttger U, Waser R, Menzel S. Impact of the Ohmic electrode on the endurance of oxide-based resistive switching memory. IEEE Transactions on Electron Devices. 2021; **68**(3):1024-1030. DOI: 10.1109/ TED.2021.3049765

[31] Wang LG, Qian X, Cao YQ, Cao ZY, Fang GY, Li AD, et al. Excellent resistive switching properties of atomic layer-deposited Al 2 O 3/HfO 2/Al 2 O 3 trilayer structures for non-volatile memory applications. Nanoscale Research Letters. 2015; **10**(1):1-8

[32] Walczyk C, Walczyk D, Schroeder T, Bertaud T, Sowinska M, Lukosius M, et al. Impact of temperature on the resistive switching behavior of embedded HfO_2-based RRAM devices. IEEE Transactions on Electron Devices. 2011;**58**(9):3124-3131

[33] Mohapatra SR, Tsuruoka T, Krishnan K, Hasegawa T, Aono M. Effects of temperature and ambient pressure on the resistive switching behaviour of polymer-based atomic switches. Journal of Materials Chemistry C. 2015;**3**(22):5715-5720

[34] Raja V, Hadiyal K, Nath AK, Viannie LR, Sonar P, Molina-Reyes J, et al. Effect of controlled humidity on resistive switching of multilayer VO2 devices. Materials Science and Engineering: B. 2021;**264**:114968

[35] Jaafar AH, Kemp NT. Wavelength dependent light tunable resistive switching graphene oxide nonvolatile memory devices. Carbon. 2019;**153**:81-88

[36] Sahu DP, Jammalamadaka SN. Remote control of resistive switching in TiO 2 based resistive random access memory device. Scientific Reports. 2017; 7(1):1-8

[37] Murtagh O, Walls B, Shvets IV. Controlling the resistive switching hysteresis in VO2 thin films via application of pulsed voltage. Applied Physics Letters. 2020;**117**(6):063501

[38] TohruTsuruoka KT, Hasegawa T, Aono M. Temperature effects on the switching kinetics of a Cu–Ta_2O_5-based atomic switch. Nanotechnology. 2011;**22**: 254013

[39] Seiyama T, Yamazoe N, Arai H. Ceramic humidity sensors. Sensors and Actuators. 1983;**4**:85-96

[40] Feng Z, El Gabaly F, Ye X, et al. Fast vacancy-mediated oxygen ion incorporation across the ceria–gas electrochemical interface. Nature Communications. 2014;**5**:4374

[41] Shirpour M et al. On the proton conductivity in pure and gadolinium doped nanocrystalline cerium oxide. Physical Chemistry Chemical Physics. 2011;**13**(3):937-940

[42] Barsan N, Weimar U. Conduction model of metal oxide gas sensors. Journal of Electroceramics. 2001;7(3):143-167

[43] Raz S, Sasaki K, Maier J, Riess I. Characterization of adsorbed water layers on Y2O3-doped ZrO2. Solid State Ionics. 2001;**143**(2):181-204

[44] Tsuruoka T, Terabe K, Hasegawa T, Valov I, Waser R, Aono M. Effects of moisture on the switching characteristics of oxide-based, gapless-type atomic switches. Advanced Functional Materials. 2012;**22**(1):70-77

[45] Messerschmitt F, Kubicek M, Rupp JLM. How does moisture affect the physical property of memristance for anionic–electronic resistive switching memories? Advanced Functional Materials. 2015;**25**(32):5117-5125

[46] Tan H, Liu G, Li R-W. Multifunctional Optoelectronic Device Based on Resistive Switching Effects in Recent Development in Optoelectronic Devices. Vol. 29. London, UK: Intech Open; 2018. p. 27

[47] Stathopoulos S, Tzouvadaki I, Prodromakis T. UV induced resistive switching in hybrid polymer metal oxide memristors. Scientific Reports. 2020; **10**(1):1-6

[48] Feng X, Liu X, Ang KW. 2D photonic memristor beyond graphene: